SHODDY

SHODDY

FROM DEVIL'S DUST TO THE RENAISSANCE OF RAGS

Hanna Rose Shell

The University of Chicago Press
Chicago and London

The University of Chicago Press, Chicago 60637
The University of Chicago Press, Ltd., London

Published 2020
Printed in the United States of America

29 28 27 26 25 24 23 22 21 20   1 2 3 4 5

ISBN-13: 978-0-226-37775-9 (cloth)
ISBN-13: 978-0-226-69822-9 (e-book)
DOI: https://doi.org/10.7208/chicago/9780226698229.001.0001

Library of Congress Cataloging-in-Publication Data

Names: Shell, Hanna Rose, author.
Title: Shoddy : from devil's dust to the renaissance of rags / Hanna Rose Shell.
Description: Chicago ; London : The University of Chicago Press, 2020. | Includes bibliographical references and index.
Identifiers: LCCN 2020001639 | ISBN 9780226377759 (cloth) | ISBN 9780226698229 (ebook)
Subjects: LCSH: Wool fabrics—History—19th century. | Textile industry—History—19th century. | Recycled products—History—19th century. | Textile industry—England—History. | Textile industry—United States—History. | Used clothing industry—History.
Classification: LCC HD9909.S59 S54 2020 | DDC 677/.312—dc23
LC record available at https://lccn.loc.gov/2020001639

♾This paper meets the requirements of ANSI/NISO Z39.48-1992 (Permanence of Paper).

For Jason and Yvette

We might almost moralise on the metempsychosis
of wool, the transfer of soul from one coat to another.
Nothing, so long as it has substantial existence,
is really and permanently useless.

"DEVIL'S DUST" (1861)

Contents

PROLOGUE

Finding Shoddy

If you have walked into the fast-fashion chain H&M recently, there will be a chant repeating on loop every seven and a half minutes. Between trashy pop songs is a deranged mantra, really: a call to shoppers to participate in a clothing donation scheme, in which preworn clothing can be exchanged in store for a discount on that day's purchase. In the name of the environment, or a general idea about how to be a good person, the call goes out. "Shoppers bring in your old, worn, or unworn clothes" to any H&M store, and we will do something good with it. In exchange you will get a 15 percent off coupon on your purchases that day. And a feeling that somehow you are not feeding directly into the very hands of the fast-fashion monster that has gotten you in the door to begin with: "*H&M Conscious*" a $2.99 reusable tote is branded.[1] Bringing clothes in puts directly into a for-profit system what has generally—in the last half century or so—been the domain of nonprofit charities. Interestingly, here the gist of the campaign is an environmental pitch—rather than one about poor people who can't afford clothes. It begins to drone, either sinking into one's psyche or driving one directly out of the store. "Shoddy" is the word left unspoken.

Shoddy is amazing.

Old Clothes Odyssey

I've been searching after shoddy for over two decades; just learning of its existence required a circuitous path. My shoddy odyssey began as part of a collaborative documentary project whose formation was the outgrowth of a lifelong fascination with old clothing. The film *Secondhand (Pèpè)* investigated the question of where the clothes we donate to Goodwill go.[2] We often consider it a great act of generosity when we decide to give clothes to such an organization, or if not deeply generous, at least a small act on behalf of recycling and the environment. There are two main reasons why we donate clothes: one, our closets are full and we need room for new clothes; two, more tellingly, it makes us feel better. We aren't simply consuming (we tell ourselves); we're giving back to our communities. We believe that the T-shirt or jacket, or whatever else we might deposit in one of those bins, will end up on a needy and worthy person's back, sparking joy in someone else's life, and not in the local landfill. And yet only a tiny fraction of donated clothing winds up in the possession of people who either buy it *as clothing* off the racks or receive it *as clothing* as charity. So what *does* happen? Where do the clothes go, and what do they become?

One day early in 2002, I found myself standing knee-deep in a textile mountain. I was at Dollar-a-Pound, surrounded by a small crowd of gleaning others, all of us on the ground at Harbor Textiles in Cambridge, Massachusetts. Here, discarded items of clothing were recast as bulk raw material measured in weight. We amassed selections from the heap, culled from the more or less (who's to know, after all) clean or dirty textile masses, what suited our eyes and fingers and caused our sense of smell little enough offense, or perhaps some small degree of charm. I'd been coming here since the early 1990s, often arriving early enough to see the wires clipped from the massive half-ton bales of compressed textile articles. At Dollar-a-Pound, I met people of all ages, backgrounds, sensibilities, and socioeconomic circumstances. Their outlooks on secondhand clothing differed, and there was a nitty-gritty singularity to each different piece they touched. Together they wove a complex tapestry before my eyes. My interest was in tracing relationships between origins and destinations of these articles, what the articles might become, and the connection between this and the lives and livelihoods of those who collect, sort, reprocess, and reuse them. And they spoke openly and shared generously their own and others' ideas about where the rags had come from, and to what use—where and to what effect—they would go. I followed

Fig. P.1 Diverse "mixed grade" used clothing compressed into a one-ton bale. Photo by Hanna Rose Shell.

them—their clothing and their stories—and witnessed as different kinds of people and different grades of rags converged (fig. P.1).

Traveling from Boston, to Port-au-Prince, to Miami, from the back rooms of the largest rag sorters in North America, to the old clothes mending stalls in Carrefour, Haiti, to an annual meeting of the Secondary Materials and Recycled Textiles (SMART) Association at the Newark Airport, it became clear that issues of human mobility and social identity were at the core of the answers I was seeking. It was just as my project was turning historical that I found shoddy. In the process of seeking context for the role of immigrant communities—Jews, in particular—in the flourishing of the British and North American old clothing industries (and related stories of "rags to riches"), I first came across shoddy, by which I mean the word *shoddy*, used differently than I'd seen it used before. The first headline that caught my eye was "Use of Shoddy Is Greatest in America: Workingmen Here Literally Wearing the World's Old Clothes," from the financial supplement to a 1904 issue of the *New York Times*.[3] And then, by way of contrast: "Only Americans Wear Virgin Wool: Order Gives Our

Soldiers Pure Worsted Uniforms, Though 'Shoddy' Satisfies Fighters of Other Nations," proclaimed a *New York Times* headline from February 10, 1918.[4]

I knew, or thought I knew, that *shoddy* meant something second-rate or badly rendered. Shoddy wasn't terrible, but it was definitely far from good: poorly, or at the very least indifferently, manufactured; "shoddy manufacture" was the first thing that popped into my head. In these headlines, however, *shoddy* was not used as an adjective; rather it was a noun, a *thing*—something the military either was or wasn't putting into army uniforms. Whatever the thing was, it compelled me. An entry for *shoddy* in the 1911 edition of the *Encyclopaedia Britannica* untangled at least a preliminary notion of the noun's referent: specifically, recycled wool or, more specifically, wool that once made into clothing, worn for years, and tossed when threadbare was resurrected and, sometimes along with wool fresh from the sheep, spun into new cloth. The *Britannica* article mentioned a mythical origin point—two towns within the so-called Rhubarb Triangle in West Yorkshire, near the River Calder, and wedged between the two major wool cities of Leeds and Bradford—and one of several apocryphal etymologies.

I learned that different sources disagree on the word's etymology; some say *shoddy* derived from an Arabic word meaning to reuse; others declare it a Yorkshirism for *should* (as in, that material should work). The possibilities continue: some are certain it came from the English word *shod*; others swore it is from the Old English *scádan*, meaning "to divide, separate"; or *shoad*, a term having to do with mining for valuables, or *shods*, beneath the surface of the earth.

A few years later, in May 2011, I found an opportunity to investigate shoddy once more. I was in London giving lectures at the Natural History Museum and King's College, and I decided to take a couple days off to take the train up to West Yorkshire. What of this "shoddy," if anything, might I find? I headed north. My destination was Batley, one of those two towns mentioned in the 1911 article. I changed trains at Manchester, and from there continued on to Batley. Not certain what to expect, I imagined arriving in *Wuthering Heights* country, a bit of William Blake's "pleasant pastures" and "mountains green." On the way, I found blustery winds over rolling hills topped with heather, low-slung farms surrounded by sheep, but also the sandstone factories and varying degrees of post-industrial decrepitude. All the features typically associated with bucolic English landscapes, in other words, but with a determined grayness always threatening to overwhelm, a bit of the aura of Blake's "dark satanic mills."[5]

West Yorkshire (a region known officially as the West Riding of Yorkshire until 1974) was transformed by the Industrial Revolution, which started in the region in the late eighteenth century, relatively soon after the establishment of the cotton industry just to the southwest in Manchester. Both coal and waterpower were in ample supply throughout the area, and it did not take long for enterprising manufacturers from farther south in the country to suspect that lessons learned in the mass processing of cotton, a business that had swelled Manchester into a city of global proportions, could rouse West Yorkshire's cottage woolen and worsted industries as well, which up to that point had operated either via the "putting-out" (through subcontracting) or the artisanal workshop models. By the early 1800s, the populations of the small villages of Batley and Dewsbury grew into sizable towns; those of Bradford, Leeds, and the other manufacturing centers of Manchester and Liverpool exploded in the years after several of the largest Enclosure Acts, including a major one in 1801, when open fields and "wastes" were closed to the peasantry and consolidated into private property.[6]

After several hours of staring out the train window, I was met at the Batley train station by a retired sportscaster named Malcolm Haigh, his eyes already aglow. "Pleasure to meet you," he said; he had heard I was interested in shoddy, and, well, he had a lot to say regarding shoddy. Malcolm had lived elsewhere in Northern England but always found himself returning to Batley (fig. P.2). Batley has been his literal and spiritual home since 1936. Malcolm looked around the town and saw wool recycling and its effects on human life everywhere; it was the place that "invented green," as he put it. Over the next several hours, he brought me around the town, making introductions and sharing his passion for reworked wool.

Batley was Malcolm's home; shoddy was his passion. Since the topic of shoddy had been on my mind for almost a decade by this point, he was amazing company. A statue erected in the center of the town square, between Batley City Hall and the Council Library, featured bolts of "renaissance" cloth and bales of rags cast in cement. In their midst was a

Fig. P.2 Looking out from Batley Train Station onto Station Road. Photo by Hanna Rose Shell.

smaller cement cast, commemorating Malcolm's self-published *History of Batley*. He looked around himself and saw the history of wool products everywhere. That the buildings in central Batley were white, Malcolm told me, well, it wasn't always so, and the reason was shoddy.

> When I was a young man, a youngster, I thought stone was black! All the smoke and the grit was impregnating all the stone in the buildings, so everything was black! I thought stone was black when I was a kid. . . . Back then I lived on the north side, and you couldn't see to the south side except at weekends, when it was just sort of perpetually misty. Then once a year, the town had a holiday, and everything closed down. And suddenly, in that week, you could see things on the other side of the valley you'd never see the rest of the year.[7]

To many of the mill workers, he recalled from his childhood, there was something thrilling about the sheer quantity of clothes their work made available, and considerable pride derived from the fact that it was being produced so close to home. Batley and its neighbor, the rival shoddy town of Dewsbury, had grown enormously in the late nineteenth century. They became places where "textile men" collected and worked on discarded leftovers from the central mills and collected rags from near and far.

Malcolm grew up during World War II and its aftermath, when the area was left decimated. Then in the 1950s, a huge influx of Asian immigrants arrived to repopulate the still-thriving textile mills in the region.[8] He remembered, from before the war, a Batley firm called Blackburn's that made high-end (shoddy) cloth that was exported to America. At the lower end of the market was a whole range of other varieties of shoddy cloth used for men's suits, women's suits, and at times a virtually endless supply of military uniforms. As he remembered it, all the lower and laboring classes in Batley, at whatever age, received a new suit—worsted—once a year. Wearing these, most of his working-class family and friends attended church at the Zion Methodist Chapel downtown (fig. P.3), with its elegant stained-glass panels each funded by a different late nineteenth-century shoddy family (plate 1). There they rubbed shoulders with shoddy factory owners at what had been baptized the "Shoddy Temple" half a century earlier.

"We didn't buy them anytime like they do now," Malcolm recalled. The town tradition was for kids to put on their new suit and then make the rounds at relatives' homes to show off their new kit. Grandparents, aunts, and uncles would dutifully *ooh* and *ah* and place a coin in the top jacket pocket. Those coins were to be later surrendered to parents and saved toward the purchase of next year's slightly too big suit. I asked Malcolm if his suits were made from shoddy. "Almost certainly," he said. "It being Batley." Three months later I returned to the Heavy Woollen District for an extended stay.

The history of shoddy, and even more so the human and nonhuman narratives surrounding it, had already compelled me sufficiently that I had decided it would be a large project. Malcolm and I again walked through the town, through the area where the old rag warehouses are, where rags were stored and sold. Most of these buildings, with elaborate designs and details on the street-side edifice, are now empty. We could see some with hooks still dangling, hooks used to lift up and move down bales of rags.

Fig. P.3 "Shoddy Temple," now the Batley Central Methodist Church. Photo by Hanna Rose Shell.

I asked him about "devil's dust," which I'd read about. Malcolm told me that many people in town had respiratory complaints.[9] Tear up woolen rags in a confined space and the air quickly fills with dust (shoddy itself is a kind of useful dust); clean or dirty, germ-ridden or not, dust is dust, and at the very least one's lungs bore the cost. "But nobody took any notice," Malcolm said. "So they just breathed it in, and of course it got into their chest, and the fact that they were smoking at the same time didn't help. So eventually lots and lots of people got bad chests, but it was accepted, it was part of the job!"

I began a series of visits to local archives and musty warehouse basements, from the Batley Public Library, to the University of Leeds Library's Special Collections. An inspired gardener invited me, with video camera and sound recorder, into his living room to show me his "shoddy weed" herbarium, and a fifth-generation shoddy manufacturer gave me the contents of the sample cabinet in the storage closet of his recently shuttered textile factory. Shoddy was everywhere. The Zion Methodist Chapel's (now the Batley Central Methodist Church) keeper, Brian Yelland, would tell me about Napoleon's blockade of Britain in the first decade of the nineteenth century. Brian saw both it and the brewing Luddite movement, many of the events of which had occurred locally, as integral to the collective understanding of the development of shoddy. Soon thereafter, I met Vivien Tomlinson, the daughter and granddaughter of successful shoddy mill owners. She took the train in from North London to meet with me. Vivien had grown up in Batley and Dewsbury, leaving the area when the industry seemed to be drying up. As we sat together amidst the rubble where a rag house had once been, she would show me several boxes subdivided into tiny square compartments, each containing a specialty sample of her family's erstwhile wares (fig. P.4).

It was the gardener John Martin—proprietor of a horticulture service called Second Nature, self-taught botanist, and "shoddy weed" collector—who took me to the shoddy heaps. And it was a rhubarb farmer by the name of Janet Oldroyd who described how the shoddy industry had been

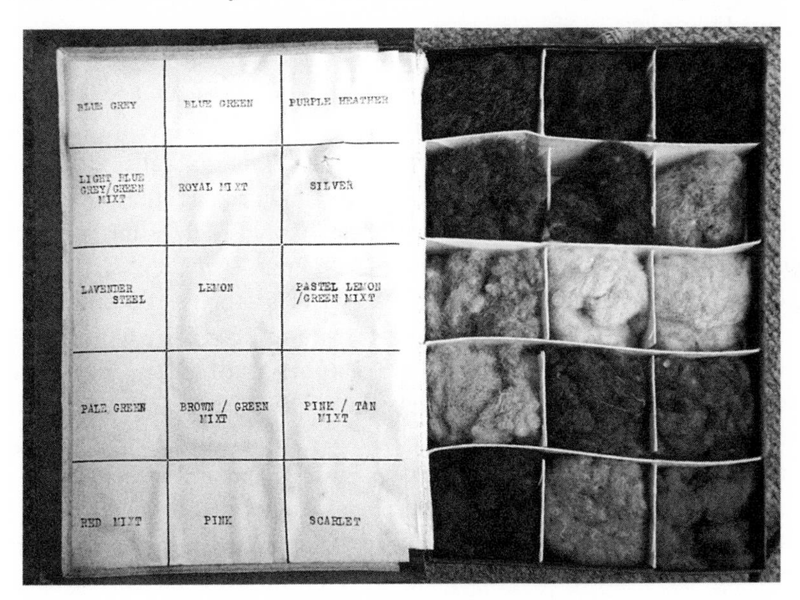

Fig. P.4 Box of shoddy samples from Vivien's collection. Photo by Hanna Rose Shell.

involved in her own family business in "forced rhubarb" production for the last century and a half. She explained how it provided her, as it had farmers throughout the region and hops farmers in Kent, a slow-acting, nitrogen-rich fertilizer.

Shoddy, as word and as thing, became more and more.

"A word is not a crystal, transparent and unchanged; it is the skin of a living thought and may vary greatly in color and content according to the circumstances and time in which it is used," wrote Supreme Court Justice Oliver Wendell Holmes in 1918.[10] And as I encountered, in addition to the evolution of its meaning as word, its manifold material, sensory, and historical facets, as well as the experiences and perspectives of these and other people and an array of historical interlocutors, shoddy became equally an evocative object and a portal into another world.

The Heap

You can see the heap as you travel between Manchester and Leeds on the M62, the motorway that crosses England's old industrial region and connects its two great port cities, Liverpool and Hull (fig. P.5). It's hard to tell what it is, though, and most travelers probably pass right by. Reaching about twenty feet at its highest point, the heap could easily be mistaken for an unusual geological formation. From a different angle, it might appear to be a mound of manure but for the fact that it's dappled gray. The smell is also different: like rotten wool or perhaps a wet dog. Up close, small items glitter amid the gray—stray rhinestones, buttons, the occasional zipper. A few peaks rise up like mini-Matterhorns (fig. P.6).

We are on the outskirts of Leeds in West Yorkshire's Heavy Woollen District, a region that in the early years of the nineteenth century emerged as a center for wool processing and a key seat of the Industrial Revolution.[11] A dusty footpath, one of many in a huge network of public pathways for ramblers, leads to the heap. At the trail's entrance is an enigmatic handmade sign: "Do not tip any more shoddy on this site." (*Tip* is a Britishism for "dump.")

Fig. P.5 The Heap. Photo by Hanna Rose Shell.

Fig. P.6 Striated Formations. Photo by Hanna Rose Shell.

This heap is composed of the shredded remains of used wool rags, socks, clothes, and remnants from the textile industry, all slowly disintegrating into the earth. Despite containing the refuse of multiple fiber-based industries, including wool scouring and rag sorting, this is not a dump in any typical sense. In various states of chemical decomposition and arranged in strata-like layers, this debris has a biological purpose; wool contains a high amount of nitrogen that it releases slowly as it breaks down, making it an excellent aid to plant growth. Here, textile waste—delivered semi-regularly, though informally, by various participants in the area's largely defunct, though nevertheless evolving, fiber-based industries—gradually turns into agricultural fertilizer that is intended for use on the surrounding fields of rhubarb, but which also feeds the bright green weeds growing around the heap (plate 2).

Today when most people hear the word *shoddy*, they think of an adjective meaning "low quality" or "badly fabricated." But, in fact, the term came into existence in the early decades of the nineteenth century as a noun, referring to a new textile material produced from old rags and tailors' clippings (plate 3).[12] Workers made it by shredding wool rags in what were christened "devils," grinding machines equipped with sharp teeth (plate 4). Recycled waste and other leftovers were turned into plentiful "new" raw materials in the "shoddy towns" of Batley and Dewsbury, just outside of Leeds. Over the next century, shoddy—along with a related textile waste derivative known as "mungo," which appeared in the mid-1830s—was widely used in the production of suits, army uniforms, slaves' clothing, carpet lining, and mattress stuffing. Leftovers from these processes landed on the fields.

Discards from wool scouring, as well as torn-up woolen rags, had been used informally before—for fertilizer and saddle stuffing, for example—but systematic processes of collection, sorting, grinding, and respinning were new.[13] "The shoddy system," as it was termed by its late Victorian promoters, reconfigured the clothing industry—and with it, machines, communities, and landscapes such as this. With its origins in clothes that were previously worn (often to pieces) by unknown and unknowable others, shoddy acted as a discomforting intermediary between human bodies and social classes. The embodiment of a complex system of materials, processes, and social structures, it became both a magnet for vitriol and a marker for the erosion—real or imagined—of boundaries between waste and manufacture, rich and poor, hand and machine.

*

Shoddy was both the product and the process by means of which used woolen garments and discarded bits from the wool-scouring process were collected and "opened" (in the language of the time), ground in a machine called a rag picker ("the devil" that produced "devil's dust"). The material was then respun for the fashioning of "new" clothes or, as it was once also known, "renaissance wool," or used as fertilizer, or stuffed into saddles and mattresses (fig. P.7).

By the early twentieth century, about a hundred years after its emergence, shoddy had become both pervasive and politically and culturally controversial on multiple levels. According to the powerful wool growers' lobby, reused wool threatened their market. To cite one example: after World War I, a surplus of army uniforms created a large surplus of wool waste. In the immediate postwar period, the wool lobby appropriated the now well-known term *virgin wool* in order to distinguish their own product from what they called the "adulterated" (i.e., "shoddy") variety.[14] The conceptualization of the "virginity" of wool—resonant today in the labels on our sweaters—thus emerged as an effort by the wool industry to counter shoddy's appeal: to make shoddy seem *shoddy* (in its modern, adjectival meaning), and to counter its positive characterization by nineteenth-century and early twentieth-century industrialists and poets alike in terms such as *regeneration*, *rebirth*, the *pinnacle of civilization*, as well as the aforementioned *renaissance* and even *resurrection*.

Partly encouraged by the wool lobby, public health experts worried about long-standing concerns surrounding sanitation and disease: How could old clothes be disinfected? How could one prevent rags brought into the United States, or from one state to the next, from carrying along the infectious diseases that afflicted previous wearers and handlers? Furthermore, what did proof of cleanliness even look like?[15]

From a contemporary vantage point, what was at stake more generally in the issues that shoddy raised? Perusal of *Weaver v. Palmer Bros. Co.*—a case in the US Supreme Court that involved old clothes, the interstate

Fig. P.7 Sample of shoddy material, and fabric produced with it, made by Henry Day & Sons, Ltd., of Dewsbury, marketed as Khaki S.D. No. 4. Image by Hanna Rose Shell.

sale of discount mattresses, and the legally contested due process clause—provides the beginning of an answer.[16] On March 8, 1926, Supreme Court Justice Oliver Wendell Holmes (fig. P.8) delivered the dissenting opinion in the case. In question was the constitutionality of a Pennsylvania law (Pa. Ls. 1923, c. 802) that banned the sale across state lines of reclaimed wool in a range of products known as so-called "comfortables." The ban applied to cushions of all shapes and sizes: mattresses for king-size beds and cribs, fuzzy toys and muffs for warming children's fingers in wintertime.[17] Palmer Brothers Co. was a Connecticut-based mattress company that had sold approximately 750,000 mattresses made of rags across state lines annually. The court struck down the ban as a violation of the due pro-

Fig. P.8 Portrait of Justice Oliver Wendell Holmes Jr., in his Supreme Court office. (Courtesy of the Library of Congress, National Photo Company Collection.)

cess clause of the Fourteenth Amendment in what was generally regarded as a victory for multiple business interests.[18] For Holmes, as well as fellow dissenter Louis Brandeis, however, the immediate issue was consumer protection; he was not convinced that consumers would be adequately informed about the risks he perceived.[19]

Suspicions about the relationship between clothing and contagion are likely to have weighed on Holmes as he considered the case. He himself had borne witness to the ragged and contaminated uniforms, most made of shoddy woolen fabric, and bandages on shattered bodies scattered across myriad Civil War battlefields. If this were not enough, his own father—the distinguished author, physician, and surgeon Oliver Wendell Holmes Sr.,—had addressed the issue of used textiles as infectious agents, decades before the development of germ theory.[20]

But there were other issues at stake, as well. In a deeper psychological and moral sense, wearing someone else's old clothes so close to one's own

SHODDY
ON THE BRAIN.

Written and composed by David A. Warden.

Send Johnson 25 cents, and he will forward you the Music of this song.

In times like these, the nation sees,
Dear friends, and not a few,
Who deal in rags, and coffee bags,
Press'd out, and dyed in glue.
These patriots strive to keep alive
The war, from Gulf to Maine,
They don't propose, the strife to close,
While Shoddy's on the Brain.
Shoddy, Shoddy, tell a body,
How you work a vein,
Shoddy, Shoddy, diamond body,
Floating on the brain.

'Tis sweet to muse upon the news,
How Shoddy at the ball,
Wore skirts of lace, with witching grace,
That fairly captur'd all.
And then THOSE pearls, 'mid Grecian curls,
Appear'd like drops of rain,
That sparkle bright, in rain-bow light,
While Shoddy swayed the brain.
Shoddy, Shoddy, &c.

When church bells ring, the coaches bring,
The pious and elite,
The vulgar walk, and perhaps would talk,

Fig. P.9 "Shoddy on the Brain." Lyrics for a song written and composed by David A. Warden and published in the early 1860s in Philadelphia by J. H. Johnson. (Courtesy of the Library of Congress, American Song Sheets, Rare Books and Special Collections.)

skin was discomforting in and of itself. Especially when that person might have been anyone or, in the case of shoddy, a motley assortment of people whose clothes had been ground together like chopped meat. As Holmes deliberated on the case in 1925, just a little over seven years after the armistice, a World War I veteran might have been sleeping on a mattress whose springs were lined and quilted subunits stuffed with a whole contingent of dead enemy soldiers' overcoats. (A coterie of European writers and poets, from Hans Christian Andersen to Charles Baudelaire, found lyrical meanings in comparable mental images in the previous century, as shown by such writings as Andersen's "The Rags" and Baudelaire's "Rag Picker's Wine.")[21] Holmes, as a decorated American Civil War veteran, is likely to have had a host of profound and mostly negative associations with shoddy well before the circuit court sent up the *Weaver v. Palmer* case. The shoddy used to make Union uniforms in that war had often been of appallingly low quality. As New England companies filled their coffers off the backs of struggling soldiers in the early 1860s, producing what were widely acknowledged to be disintegrating uniforms and asthmatic blankets, *shoddy* the noun was increasingly used as an adjective that captured a host of personal, ethical, commercial, and societal failings (fig. P.9).

And at bottom, there was an epistemological issue: how to understand what shoddy even is, or was, posed a philosophical as well as a practical problem. Shoddy was, in its essence, unknowable in terms of what would assuage fears about its intimate association with unknown others, as a vector for disease, and as the product of corrupt and venal manufacturers. In the years preceding *Weaver v. Palmer Bros.*, deliberations for which lasted the final three months of 1925, Palmer Bros. of Connecticut had been shipping hundreds of thousands of shoddy mattresses into Pennsylvania, to the chagrin of in-state manufacturers and wool growers. Both groups also charged shoddy companies with deception insofar as the shoddy mattresses were not identified as such. (The ominous and still omnipresent labels that read "Do not remove this tag under penalty of law," alongside itemized fiber contents, were mandated subsequent to this case and in response to the issues therein raised.)[22] To be sure, Palmer Bros. claimed to sterilize its materials using a steam process shown to be effective for that purpose in the previous century. But, asked Holmes, who is to know whether sterilization is in fact performed? Could the steam penetrate the seemingly impenetrable giant bales in which the old clothes were shipped from overseas? And even if this *could* be determined, could one be certain that the specific shoddy batting in a given mattress was sterile, without destroying the mattress in the process? As Holmes himself put it, "It is admitted [on all sides] to be impossible to distinguish the innocent from the infected product in any practicable way, when it is made up into the comfortables."[23]

Deeper fears, often stated in explicitly moral terms, surfaced in the statement to the Supreme Court made by the legal representative of the appellant, the attorney general of Pennsylvania:

> The evil [prior to the ban] was the insanitary condition that existed in the bedding industry, and the insanitary product, which was coming into the hands of the consuming public, as well as the fraud, and deception which was being practiced in the make-up of the articles sold. Much knowledge of this evil was and is a part of the common knowledge of mankind. . . . In

> the process of manufacture [shoddy] loses its identity. Its nature facilitates the practice of fraud and deceit.[24]

The stakes were thus high and multifarious; shoddy was a substance through which a debate about old clothes and mattress stuffing could be cast in terms of a battle between good and evil, order and dissolution, staged in the nation's highest court. "The prohibition of the use of shoddy, new or old, even when sterilized, is unreasonable and arbitrary," the court decided, but for Holmes and the other dissenters, this seemed far from true.

These issues point to the ways in which shoddy (which "loses its identity," in Holmes's words, in the very "process of its manufacture"), a term whose original nominative meaning is today all but forgotten, generated such panic and alarm.

Shoddy's manufacture requires, in part, the active fashioning of erasure—of making disappear the sick people, old clothes, and their wearers and handlers, the physical and emotional spaces where they came from, the people who were touching and collecting them. Its instability as a commodity results from both the constantly shifting origins of its component rags and also the inherent imperfection of such attempts at erasure; a trace will always remain. This book's purpose is not only to challenge and provoke, but also to reveal the hidden graces of the shoddy world, beauties both terrible and tender. Three acts and an epilogue cross axes of chronology and geography. Taken together, they open up the rich world of this peculiarly unstable entity in hopes that a fuller understanding of that world can help us think anew about the myriad textile surfaces that both separate and link the human and the nonhuman, the human-made and what is itself composed of human stuff, the "alien" and the "comfortingly" familiar.

ACT I

Devil's Dust

Day after day, I must thatch myself anew; day after day, this despicable thatch must lose some film of its thickness; some film of it, frayed away by tear and wear, must be brushed-off into the Ashpit, into the Laystall; till by degrees the whole has been brushed thither, and I, the dust-making, patent Rag-grinder, get new material to grind down.

THOMAS CARLYLE, *SARTOR RESARTUS* (1833–36)

For as long as there have been clothes ("worn" by definition), there have been old clothes. And certainly, long before shoddy, there was a trade in textile reuse. From their origins in the first days of human culture, clothes were portable technologies, wearable media, and temporary prostheses, shaped by the demands of a mobile body and inscribed with markers of that body's history. They were (and remain) artifacts—articles of manufacture—in continual flux, though in earlier times exchanges of old and new were fewer and farther between.

Throughout the Middle Ages and the Renaissance, clothing, once shaped to a given body, might be worn for years, sometimes a lifetime. Through the movements of a body in time, its clothes would acquire increasingly personal and human characteristics—worn knees and elbows, a stretched waist. Stains, patches, tears, and color changes accompanied a life journey, or at least several decades thereof. In the Renaissance, it was common for servants to sell their masters' old clothing to peasants in neighboring villages. In the seventeenth and eighteenth centuries, itinerant rag and old clothes dealing grew into a profession, increasingly associated with the manufacture of paper, including paper money. In particular, the dealer became an intermediary between wearers, marking a transitional phase in an article's mobile life history (fig. 1.1).

But with the advent of the new forms of production characteristic of the Industrial Revolution, the pace and volume of manufacture skyrocketed. And the more clothes there were, the more they changed hands. Over the coming century, mechanical looms, spinning mules, and sewing machines would make clothing easier to produce and more affordable to populations, if also at times less durable. With more clothes for the taking, the working and middle class purchased more, therefore wearing each garment for less and less time and creating more waste material, or "clippings," in the production process.[1]

New clothes became common instead of rare, and worn clothes were discarded or exchanged at an ever-quicker pace. As the paths of clothing increasingly departed from the bodies of single individuals and their families, they began to travel farther and farther. The result of excess, among other things, was a new genre of, and set of possibilities for, salvaged clothing. It remained a good to be resold as is, in a system of grassroots circulation, but also a medium for transformation and exchange. Textile media—both the material out of which technologies of wear are produced and the means by which use and meaning are transmitted by them—would evolve into the "rag and shoddy system," to use the terminology of a nineteenth-century West Yorkshire mill owner turned historian. In the course of that development, embodied ritual was partly subsumed by industrialized anonymity—creating a new resource, to be sure, but one littered with traces of unknowable people, obscured origins.[2]

Shoddy emerged within this sociotechnical context as fabric material manufactured from new wool spun together with the shredded fibers of old or recycled wool. Shoddy, in its original meaning, is a combination of used wool, often formerly worn for years and discarded when threadbare, with a smaller amount of new wool fresh from the sheep and thereby resurrected as "new" yarn or cloth. Sources, both historical and contemporary (the latter being interview subjects in twenty-first century West Yorkshire) disagree on the etymology of the word *shoddy*, along with that of shoddy's finer cousin, *mungo*.[3]

Industrial-scale recycling was born, in part, through the development of specially purposed machinery for the sorting, grinding, scouring, and

Fig. 1.1 Ralph Sly-boots. Says Roger Hatchet-face to Ralph Sly-boots, the old Rag man: ". . . Here!—you that cry old Cloaks and Suits, / Come hither Friend—what do you think, / Of half a Crown, to make you drink?" The ragman responds: "I think I cannot better do, / Then take it Sir—and thank ye too." (George Arents Collection, New York Public Library.)

baling of old, used wool. Recycled waste and other residue were turned into plentiful "new" raw materials in what came to be known as the "shoddy towns." These mill towns were located within the geographic triangle formed by the West Yorkshire cities of Bradford, Leeds, and Wakefield, and the two most important of these were Batley and Dewsbury (fig. 1.2). By the second decade of the nineteenth century, human workers shredded wool rags in the sharp teeth of the rag-grinding machines colloquially known as "devils." A couple decades later, the shoddy industry was entrenched.[4]

Among those setting the new intellectual tone of the era was Thomas Carlyle, an influential Scottish literary critic, philosopher and historian. Carlyle's "Signs of the Times" appeared in 1829. In that essay, Carlyle (who had previously written literary reviews and had been greatly influenced by the philosophical and literary works of the German Idealists and Romantics, including Goethe) applied his critical faculty to the economic and social changes currently at work in the United Kingdom, introducing the phrases "the age of machinery" and "mechanical age." In calling the age mechanical, he meant to refer not only to the many new products of the evolving industrial arts, but also to new systems of bureaucracy and modes for the organization of labor.[5] Carlyle's new age encompassed a new process of dehumanization he saw as being connected to the ways money was now amassed, in which "nothing is now done directly, or by hand." Carlyle laid out two distinct but complementary notions of machine: the "outward" and the "inward."[6] The "outward" form referred both to the "machine/mechanism" itself and to the social and bureaucratic system forming around and through them:

> On every hand, the living artisan is driven from his workshop, to make room for a speedier, inanimate one. The shuttle drops from the fingers of the weaver, and falls into iron fingers that ply it faster.[7]

For Carlyle, the exemplary mechanical object is the flying shuttle, which, after initially being operated by human weavers, could then be mechanized and incorporated into a power loom. The living artisan is driven from his habitus as the horse is stripped of his harness, becoming an inanimate artisan, and in the process no artisan at all. As a result, the

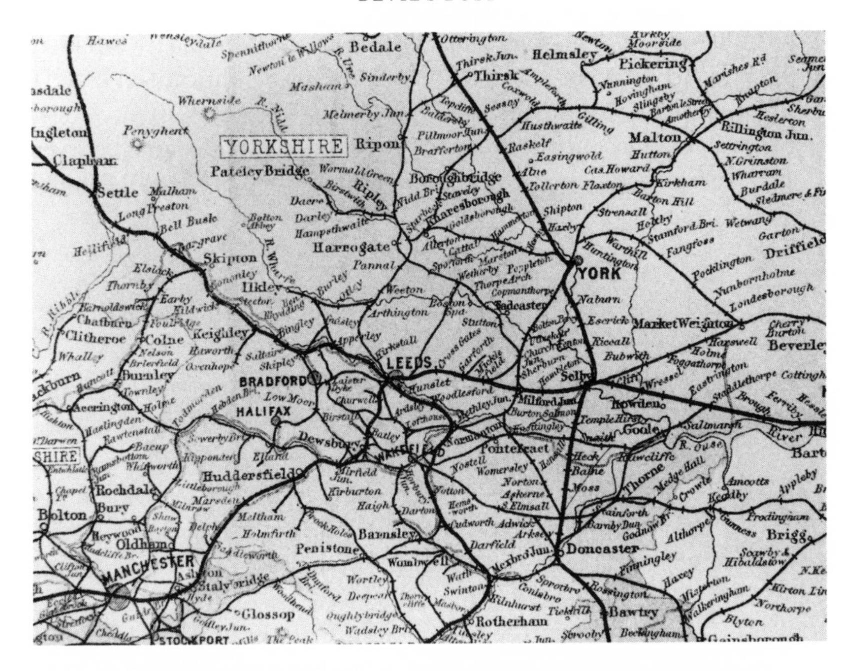

Fig. 1.2 Detail from the "New Railway Map of the British Isles," by W. H. Smith and Sons, c. 1890, showing Yorkshire district. The Heavy Woollen District exists in the triangle formed by the cities Bradford, Leeds, and Wakefield. Batley is approximately equidistant between Bradford and Wakefield, with Dewsbury one train stop south from Batley. (Courtesy of the Harvard Map Collection.)

"inner" or "spiritual" element is lost, or rather replaced by the "heart" of the machine, evacuating the very concept of the self as previously understood, as well as opening up new areas of political and economic cleavage.

Carlyle's concept of the mechanical age interpreted in the context of his later *Sartor Resartus* (1836) provides us with a way of understanding shoddy's richness as a substrate for making both meaning and modernity. This in turn provides us with a way of understanding what might otherwise be puzzling—whether it be crowds gathered by rags on the train tracks or a few especially difficult passages from the volumes of Karl Marx's *Das Kapital*, as well as passages from the work of his colleague and collaborator, Friedrich Engels. Shoddy, in fact, provides the material form for Marx's paradox, as well as a seat for British prime minister Benjamin

Disraeli's angst over his own and his nation's straddling between identities, what he called its "two nations." In multiple political and literary modalities, shoddy came to materialize the contradictions of modernity itself: its simultaneous erosion of and reinscription of boundaries between human and nature, between hand and machine, and between the found and the fabricated.

Emergence of an Industry

Shoddy emerged in the nineteenth century as both material product and industrial process. Conditions that made this possible included the preexisting infrastructure of the conventional wool industry and rag-based paper manufacturing, as well as the development of railroads and the shipping industry to move goods nationally and internationally. A turn-of-the-century wartime shortage of raw materials for industrial use coincided with shifting ideas about the nature of waste.

The dramatic growth of the West Yorkshire woolen and worsted industry at the turn of the nineteenth century, and the accompanying ballooning of the region's human population, coincided with a major influx of raw materials.[8] Wool initially came mostly from England and other parts of Great Britain, though finer wool was often imported from Italy and elsewhere. But as production capacity increased at the end of the eighteenth century, more raw materials were needed; trade with foreign countries exploded.[9] Entire fleeces and shorn wool arrived from the colony of New South Wales (Australia and New Zealand) and South America.[10] Bradford grew into the mill city of "Worstedopolis" known for the production of broadcloth and worsteds. Woolen and worsted refer to two different types of yarn, and by extension the fabrics woven with them. Woolen yarn (spun from carded wool) is relatively loose and tends to be used for making knits such as sweaters and scarves. Worsted yarn (spun from combed wool) is tighter and stronger; it tends to be used for tailored garments such as coats, suits, skirts, and blazers. (Shoddy could theoretically be used in the production of either kind of yarn.) Nearby, Leeds, already a center for woolens manufacture and the wool trade, continued to grow into a global hub for wool exchange. It was in this context that Napoleon's blockade on the English Channel would ultimately do so much economic damage.[11]

Wool and other textile waste had of course existed and been reused before the nineteenth century and the emergence of shoddy. Cotton and linen rags, for example, had been used to make individual sheets of writing paper for thousands of years.[12] And prior to the industrial era, leftover woolen materials from the scouring and carding stages, as well as vegetable matter and inorganic debris shaken out of them, would have ended up on the floor of the cottage, piled up in heaps, and eventually dumped outside haphazardly, or more purposefully in heaps as garden fertilizer.[13] In that era, cloth left over after a garment was completed might be used to patch something else, or possibly be bound up as stuffing for furniture or horse saddles.[14] These pathways were often informal, rooted in handwork, personal eccentricity, and a piecemeal "make do" approach to household economy.[15] Alternately, they became part of semi-formalized economies organized at multiple scales for the reuse of clothing as clothing—sometimes purchased, sometimes stolen, sometimes bartered—for immediate use or resale.[16] But the rapidly accelerating pace of wool manufacture, combined with its growing centralization into mill-based systems of production, created new volumes of industrial waste.

As the volume of textile waste increased exponentially, its distribution across the landscape shifted.[17] Textile waste no longer ended up in small backyard piles. Instead it built up in urban factory settings, far from any one individual's household garden or mending basket. Every key stage in the production process—scouring, carding, spinning, and weaving—created some amount of refuse, then potentially available for reuse. Even beyond these, the actual cutting of fabric into patterns and sewing into garments, once scaled up in the factory system, created so-called "tailors' clippings." These leftover bits and pieces of fabric (threads, linings, and other scraps) joined the textile heaps piling up in all the various facilities.

An abundance of discarded cloth at a time of high demand for woolen goods during the period of the Napoleonic Wars created an environment well suited to the development and flourishing of the textile recycling industry.[18] The new material was drawn from three different kinds of wool waste: scouring refuse, tailors' clippings, and most significantly bales upon bales of shredded old wool rags—all of which powered the salvage technology that came to be known as "shoddy manufacture." Machines for carding new wool could be slightly adapted to process old soft rags. The rags were collected initially from the region, but soon from throughout England, Scotland, and Ireland, and then continental Europe. Once sorted, the various kinds of fabrics would be shredded in the rag-picking machinery. Broken down to its elemental fibrous state, the shoddy wool could then be respun (and with a smaller or larger percentage of "new" wool incorporated) into dramatically less expensive fabric for use in a range of garments and blankets.[19] An emergent system of salvage technology collected discarded rags, sorted them, incorporated them into shoddy garments—red coats and blue pants for navy uniforms, ladies' coats, blankets, carpets, and so forth.[20]

Narratives of Transmutation, Myths of Invention

However shoddy came to be, whoever ground it first, there is a local story often told. The development of the process of shoddy manufacture is, in Batley at least, credited to a native son, a young Batley mill owner and clothier named Benjamin Law, whose production of the first shoddy is dated to 1813 (fig. 1.3).[21] The story often told about his innovation is that Law traveled to London to scour the markets for a good price on wool. While there, he came across a saddler using torn woolen rags, formerly old sweaters and scarves, to stuff saddles, which he then sold alongside other equine equipment.[22] The traveling clothier picked up a piece of stuffing, and to his surprise there was still some "give" in the wool; he could twist and pull and it didn't break. It occurred to Law that torn woolen rags could be further shredded and then respun into a kind of "renaissance" yarn, which could be woven into a new type of fabric: a more economical incarnation of wool.[23]

When Law returned to Batley from London, he and his brother-in-law, Ben Parr, began to stockpile discarded clothes, both those collected by ragpickers for resale as clothes and rags that would otherwise have been scattered on the fields as fertilizer.[24] Meanwhile, they visited an operational wool scouring and carding facility, as well as a separate mill where

Fig. 1.3 Plaque outside Batley Library. Photo by Hanna Rose Shell.

cotton and linen rags were shredded down to pulp for paper. During their visits, they tried committing to memory the design of the machines. Law hoped that, with some modifications, they could shred woolen rags into threads that would be easier to mix with virgin wool. The first pieces of cloth woven from a blend of new wool and shredded old wool were full of imperfections. Parr (whose background was in machining) determined that, in principle, two "swifts"—cylinders studded with a metal comb—could, when revolving against each other, tear up rags fed into where the two swifts met. But it was hard to get the tension right. Too loose, and the rags did not tear. Too much tension, and the combs jammed. Law then consulted a blacksmith, who suggested they replace the two-swift mechanism with one large swift surrounded by several smaller swifts, and upright conical "teeth" instead of combs. In the basic version, a cylinder covered with wooden lags, with close-set teeth, metal pins, or a combination of the two inserted. This worked better and resulted in the first

Fig. 1.4 An early model of a rag picker machine (a "rag devil"). Illustration by Robert Buehler.

machine dedicated to the tearing, shredding, and near pulverization of woolen rags; it was called a "rag picker" because it picked apart old clothes, though it would quickly earn the popular moniker "the devil" (fig. 1.4).[25]

The mechanism, and even the outward casing, has remained surprisingly constant over the past two centuries. A modern-day business owner described the machine and its role:

> We buy rags, a forty-foot trailer every two weeks, and then I've got a warehouse where I store it. We bring eight or nine bales into the factory daily. We feed it to our rag machine, that actually shreds the rags—once it's been pulled, it goes through a second operation when it runs through the carding line.

"I can show it," he continued:

> This is our carding line, it's fed in from this end, it goes through, all these are jagged teeth and fine teeth on them, and it's making finer, as it goes through it's making it finer and finer. By the time it gets to the large swift, then it's more or less ready to come through to our end of the product.[26]

In recent years, as the towns of Batley and its neighbor Dewsbury have struggled to reinvent themselves in a largely post-industrial context, they have worked to brand themselves by invoking renaissance, recycling, and variations on "being green" and through shoddy's historical incarnation by Law and its being put to paper by historian Samuel Jubb. Dissenters have piped in as well, disputing the characters, while sticking with the account of the processes involved in shoddy's origins. The development of shoddy, both invention and innovation, was two-pronged. First came the prototyping and refinement of a rag picker through modification of existing machinery.[27] Second was the organization of a labor system for the processing of old and waste wool to make finished clothes. As part of the functioning of the latter, a specialty in skilled human labor was required: the rag sorter (fig. 1.5). Unlike many other aspects of this and other parts of the textile industry as it developed over the century, rag sorting was—and remains even to this day—a job necessarily performed by flesh-and-blood human beings. The goal was to sort rags, as quickly as possible, into "grades" by color, fabric quality, state of disrepair, and so on.[28] Touch, smell, actions such as rubbing a cloth against itself by pinching between thumb and forefinger: these were all the kinds of gestures that helped the sorter work efficiently and the shoddy industry to maximize its potential.[29]

The sorter was a kind of material classification specialist, operating with the same intuition and skill sets that had characterized sorters for paper manufacture in the preceding century, placing fragments into different crates depending on material, color, and various degrees of fineness and other distinctions of quality. A retired textile sorter reflected on her experience: "Oh, if you were in the rag mill, everybody would go in pockets and find money and watches and all sort of stuff; everybody looked in the pockets to see what were in them, like it was part of the job."[30] A constant dealing with novelty lent itself to the enactment of a kind of natural history classification built largely out of what philosopher Michael Polanyi has called tacit knowledge.[31]

Fig. 1.5 James McNeill Whistler, *The Rag Gatherers*, etching and drypoint, 1858. (Metropolitan Museum of Art, Harris Brisbane Dick Fund, 1917.)

The above is most certainly true, in terms of what shoddy actually became for the region, in terms of employees, in terms of the technology, and as a model for industrial recycling more generally in the nineteenth century and beyond. Indeed the modern-day "devil," often referred to as a "garnett," works much the same as the early nineteenth-century version. As to origins, however, the "commemorative" story of Benjamin Law, as a small-time weaver in a tiny Yorkshire town who made it big, is as much a founding myth as it is rooted in, or approximates, historical accuracy. The complexities of a relationship between Law and Parr are only one, albeit one commonly shared and celebrated, means of accounting for shoddy's origins—shoddy as *word*, as *material*, and as *technology*, to say nothing of it as the identity of both an industry and the individuality of those working in its midst. In 1860 it is held that Batley produced approximately 7,000 tons of shoddy annually, with 80 firms and 550 rag sorters. By 1880, long after the shoddy industry had made the Heavy Woollen District prosperous and populous, an engineer, chemist, and shoddy fertilizer manufacturer (treating wool waste so as to facilitate its use as nitrogenous compound in farming) challenged the narrative. Ferrar Fenton wrote into the *Batley News* in 1880, and in a piece published on December 1, he shared the following account of a meeting with a group of London merchants at which

> someone mentioned that I came from Yorkshire and at once the talk ran on shoddy. I was asked who invented it. I replied it was attributed by some to a gentleman named Law, of Batley and by others to a Mr. Parr. An elderly Jewish gentleman told me I was mistaken, neither Law nor Parr invented it, they adopted it.[32]

Fenton adds a different story then, one inflected by the long-standing association between Jews, old clothes, and the refrain of "rags to riches," The concept and its putting into practice seem in this version to emerge surreptitiously in London, amidst Napoleon's blockade of 1806 and 1807, hence several years prior to Law's apocryphal triumph.

> My father was the first man who ever thought of, or made it. My father told me thus: He was an extensive dealer in second hand clothes in Whitechapel here, and when the Peninsular War broke out, after old Bonaparte had overrun Spain, owing to the stoppage of the supply of Spanish wool and the brisk demand for army goods for Sir John Moore's expedition to Spain, Spanish wool, which was then used for making them, rose to a tremendous price and my father, who as a "clothes duffer," had been used to seeing the lot of woollen dust or "mill puff" as it was then called pulled off old clothes in preparing them for market, got a new idea into his head. It occurred to him, he said, that if he bought wool in London, opened the bales, and inserted among the fleeces 50 per cent of old blankets or white

> flannels, torn up with curry combs, it would be a paying speculation. He flung aside second hand coats and trousers and covered his "doffing cushions" with old blankets and sent off a lot of doctored bales and realised full prices in Yorkshire. After this he kept at it till no more white flannels were to be got in London and when wool fell in price he turned to making flocks for stuffing saddles and mattresses to replace wool, which had previously been used for these purposes. . . . For our acute Jew must have made his first trial of mixing it with his Spanish wool about 1805–6 and an examination of the wool market quotations of 1805–6 might almost fix the month when he began it. It would also here appear that there were many involuntary users of "shoddy" at the very opening of the century. The word shoddy itself is, I believe of Hebrew origin. I would give you the root but have not my lexicon at hand. However, in a cognate language, the Perso-Arabic "shodjy" or "shodjah" signifies a tangled mass of something like "sedge-wool" and other allied ideas. This fact tends to confirm the accuracy of the statement of the gentleman that I relate above and no doubt brings me to the very root of the origin of the article that next to cotton has made one of the greatest revolutions in the textile industry.[33]

Sir George Head, writing in the mid-1830s, provides what may be the closest thing we have to an actual contemporary description; the town of Dewsbury, Batley's neighbor, he said, was known not only for its blankets but also for a "novel business that has [recently] sprung up in England," in addition to "arts and sciences," namely: "that of grinding old garments new;—literally tearing in pieces fusty old rags, collected from Scotland, Ireland, and the Continent, by a machine called a 'devil,' till a substance very like the original is reproduced. This, by the help of a small addition of new wool, is respun and manufactured into sundry useful coarse articles."[34] Capitalizing on the theme of alchemic "transmutation," Head continues:

> The trade or occupation of the late owner, his life and habits, or the filth and antiquity of the garment itself, oppose no bar to this wonderful regeneration; whether from the scarecrow or the gibbet, it makes no difference; so that, according to the change of human affairs, it no doubt frequently does happen, without figure of speech or metaphor, that the identical

> garment to-day exposed to the sun and rain of a Kentish cherry orchard or saturated with tobacco smoke on the back of a beggar in a pothouse, is doomed in its turn, "perfusus liquidis odoribus," to grace the swelling collar, or add dignified proportion to the chest of a dandy. Old-flannel petticoats, serge, and bunting, are not only unraveled and brought to their original thread by the claws of the devil, by the way, simply a series of cylinders armed with iron hooks, effectively, it is said, pulls to pieces and separates the pitch-mark of the sheep's back—which latter operation really is a job worthy of the devil himself.[35]

Shoddy's first official historian, Samuel Jubb, himself a member of a major shoddy family in Batley, would later repeat the alchemic metaphor, observing: "To the uninitiated, it must be surprising to see the rags suddenly transformed into fibrous wool; and it is in this process of grinding, that the apparent impossibility of making old rags into new cloth vanishes away."[36]

By 1830 hundreds of sorters, mostly women and children, were employed in at least thirty independent shoddy grinding facilities in Batley and Dewsbury, all clustered within a couple of miles. Within decades of its introduction, the shoddy industry spread from Northern England across the Atlantic Ocean to the states of New England. Textile workers (often their children) traveled to New England starting in the 1820s. There, they manufactured rag grinders for use by already-established spinning facilities or went on to facilitate the establishment of shoddy mills near the textile centers in Rhode Island, Massachusetts, and Connecticut. Shoddy mill owners who stayed in West Yorkshire would go on to expand into Germany and Holland, already origin points for many of the rags landing by ship and by rail at the Batley and Dewsbury auction blocks.[37]

The so-called "shoddy system" worked its way into communities, factories, and landscapes, as well as individuals' closets and carpets. As the infrastructure continued to develop in the 1830s and 1840s, the towns grew larger, and the product found its way into increasingly varied types of clothing and other textile products. By the 1840s, these formerly unknown towns had acquired a presence throughout England and on the Continent. Shoddy dealers traveled down to London to oversee the collection and transport of the capital city's rags up to Yorkshire; this became a regular part of the more general trade in secondhand goods already flourishing in European cities.[38] The keen urban observer Henry Mayhew described the network of textile materials and practices, both to which shoddy contributed, and from which it was an outcome, in a London marketplace in the 1840s:

> The old coats and trowsers are wanted for the slop-shops; they are to be "turned," and made up into new garments. The best black suits are to be "clobbered" up—and those which are more worn in parts are to be cut up and made into new cloth caps for young gentlemen, or gaiters for poor curates; whilst others are to be transformed into the "best boys' tunics." Such as are too far gone are bought to be torn to pieces by the "devil," and made up into new cloth—or "shoddy" as it is termed—while such as have already done this duty are sold for manure for the ground.[39]

Through the medium of shoddy, woolen fibers crossed the nation and moved between different physical manifestations. The shoddy substrate generally required dusting and carding, before a blending of wools, generally some proportion of new fibers entering as part of the final "shoddy blend." As an entity both substantial and apparently amorphous, shoddy could be draped and redraped on human and other bodies.

Over time, shoddy found its way into increasingly varied types of clothing and other textile products produced in England, the United States, and the Continent.[40] The arrival of railroad lines in the towns of Dewsbury and Batley made it increasingly easy for rags to come from throughout the world and for shoddy products to depart. Rag merchants held massive rag auctions for shoddy grinding factory owners next to the railroad depots. A journalist described dealers overcome by "visions of filthy rags being transmitted into shining gold" at the train depots of Batley and Dewsbury, as the thousands of bales arrived at the station from abroad.[41]

What had once been a region of sleepy villages adjacent to the major centers of Leeds and Bradford turned into its own economic and technological center and site of general interest. The *Westminster Review*, a quarterly journal founded by Jeremy Bentham, published an essay on Yorkshire, discussing at some length the rivalry between the two tradi-

Fig. 1.6 Touching the Devil. Photo by Hanna Rose Shell.

tional clothing district capitals of Bradford and Leeds in midcentury. The article highlighted the shoddy towns, describing Batley as

> the chief seat of that great latter staple of England—shoddy. This is the famous rag-capital, the tatter-metropolis, whither every beggar in Europe sends his cast-off clothes to be made into sham broadcloth for cheap gentility. Of moth-eaten coats, frowsy jackets, reechy linen, effusive cotton, and old worsted stockings—this is the last destination. Reduced to filament and a greasy pulp, by mighty tooth cylinders, the much-vexed fabrics re-enter life in the most brilliant forms.[42]

This profile captures both the promise and the criticism leveled at the "mighty tooth cylinders" of the shoddy industry. Meanwhile, the material itself increasingly became both a vector for and a materialization of interrelated political, economic, and ethical quandaries; a touching of the devil, as it were (fig. 1.6).

Devil's Dust Politics

The notion that the shoddy machinery was somehow morally suspect proved hard to quell. Widely read early observers like Head[43] had gone out of their way to emphasize certain repulsive physical features of the enterprise—its malodorous and unhealthy atmosphere, for example, and the masklike dust it left behind on its participants. As Head archly noted:

> I know not whether it be a physical or a metaphysical question whether a smell be, *de jure*, a noun and the name of a thing, having substance and dimensions, or whether it be an ethereal essence devoid of material particles,—as it were the benediction of animal matter departing from the physical to the metaphysical world, and at that very moment of its critical existence and non-existence, when it belongs to neither. . . . Some little notion may probably be given [of its repulsive and at the same time transformative character], by stating that the boys and girls who attend the mill are not only involved all the time it works in a thick cloud, so as to be hardly visible, but, whenever they emerge, appeared covered from head to foot with downy particles that entirely obscure their features, and render them in appearance like so many brown moths.[44]

Of additional concern were the industrial rag picker's violent motions, and the fact that its principal part looked like nothing so much as a set of sharp teeth. The author of a piece called "Devil's Dust," published in *Chambers's Journal* in 1861, described his impression of the machinery: "The principal part of the rag-wool machine is the *swift*, a frame provided with ten or twelve thousand vicious-looking teeth, and that rotates six or seven hundred times a minute." Wool input is "not merely *torn*; it is almost *ground*. . . . What would be the fate of James's coat, or Alphonse's jacket, when exposed to the action of such a monster, the reader may readily imagine."[45] The rag grinder increasingly came to epitomize a more general horror associated with machines.[46] The shoddy industry became emblematic of the widespread linkage between machinery and a process of dehumanization thought to be inherent in an economy rapidly shifting toward industrialization.[47] Shoddy seemed to have no respect for boundaries, making it hard to tell the pure from the substitute, the derivative from the fraudulent. Devil's dust became the encapsulation of its fraught nature.

For historical actors as distinct as a conservative member of Parliament (MP) and a young and increasingly socially conscious Friedrich Engels, and in the context of two major political issues of the 1840s, shoddy served as a potent materialization of corruption and deceit perceived as inherent in the industrial system. At issue in the early 1840s were the fates of the Factory Act (whose passage would create new stipulations on labor conditions) and the Corn Laws (protectionist measures implemented in the 1820s designed to inflate the value of crops grown on English soil). Politicians representing the interests of factory owners and the mercantile class argued for the repeal of the Corn Laws; their argument, and that of the Anti-Corn Law League more generally, was that eliminating the steep taxation on imported grains would improve the welfare of the working class by making their sustenance more affordable. The landed gentry, for whom high crop prices meant profits, vociferously opposed a repeal.[48] Conservative landowners sought to blame the factory owners, and the Factory Act was put before Parliament to offer a measure of protection.

Among the former was William Busfield Ferrand, a member of Parliament representing a substantial region of West Yorkshire in the House of Commons. Known as an irascible though charismatic speaker, Ferrand presented the shoddy trade as an especially vivid example of the ills of British manufacture, using it as a way to deflect popular animosity away from traditional landowners. Shoddy in the singular form of devil's dust became a lightning rod in arguments over the Corn Laws, with industrialists and other free-traders favoring repeal and land-owning aristocrats and democratic "Chartists" opposing it. In February 1842, Ferrand, opposing repeal, opened up the fifth day of deliberations in the House of Commons by describing

> the process which is adopted by certain manufacturers, of buying up all the old rags they can obtain, which are torn into pieces by a machine, thus converted into a kind of dust, and are then mixed with wool, which is eventually manufactured into cloth. This dust, from its nauseous nature, and from its engendering numerous diseases, has been christened by the manufacturers and workpeople of Yorkshire the "Devil's dust."[49]

Ferrand claimed to trace the term devil's dust to another, older fraudulent practice involving the manufacture of textiles (in this case, silk), thereby linking it with deceit and fraud more generally, an association repeated in later dictionary entries on the term.[50] Devil's dust both poisoned the workers' lungs, Ferrand claimed, and adulterated their cloths when used in paste to add deceptive bulk.[51] Later in the same session, he read from a letter he had received from an unnamed shoddy mill worker: "Things are worse and worse in Huddersfield; and it seems that all is over without any hope. . . . I wish you could get a full account of this shoddy trade: it is monstrous" (fig. 1.7).[52] Ferrand, credited with having introduced the terms devil's dust and the devil (in reference to rag-picking machinery) into usage outside of West Yorkshire, pointed to shoddy as not only responsible for a loss of faith in English goods,[53] but also as symbolizing the moral depravity and deceit of the manufacturing and industrial class.[54]

The response of William Fox—later reported verbatim in Francis W. Hirst's influential defense of classical political economy in *Free Trade*

Fig. 1.7 "It is monstrous." Photo by Hanna Rose Shell.

and Other Fundamental Doctrines of the Manchester School—is typical of members of the anti-protectionist "League"; far from gaining support from employees of the shoddy mills, such diatribes against repeal were opposed, Fox claimed, by the workers themselves:

> Protection! Why, what should we protect? Not a losing trade, for that is taxing all the community for the advantage of one class. . . . It would be worth your while to send a deputation down into the North. . . . You should see the multitudes flocking together in these districts, men, women, and children, persons of all ranks and classes, as to a work that calls forth the deepest sympathies of human nature. Yes, you should see them coming and mingling together in the same assembly—masters and men pouring out from the same factories. There is no heed paid there to the calumnies and stories which are circulated in some quarters; there are no symptoms there of the tyranny which has been talked of elsewhere . . . ; but there come the operatives from the factories, not choked with "devil's dust," as Mr. Ferrand says, but ready to "down with their own *dust*"[55] in the cause [of free trade].[56]

Critics and defenders of the Corn Laws alternatively invoked the title "Dewsbury Devil" in facetious equation of the rhetoric of their opponents with the noise and dust raised by the infamous machine, as in this entry from *Common Sense, or Every-Body's Magazine*, a High Church publication eager to point out the link between defense of free trade and the dissenting religious views that frequently accompanied it. After citing the letter of one such "Reverend," who had attempted to lampoon his adversaries with that title, the author claims to find a "very pretty moral in the story (for it is no fable) of the 'Dewsbury Devil and the Dissenting Preacher,'" who are indeed a "very *religious* pair—at least so far as hostility to the Corn Laws is *religion*."[57]

In sum, by the mid-1840s, devil's dust had become an emblem and rhetorical crux for competing political, economic, and religious views across a variety of overlapping spectra. Younger Tories such as Benjamin Disraeli (who first stood for election as a member of that party in 1835 and would ultimately serve as prime minister in 1868 and again from 1874 to 1880) increasingly made common cause with the Chartists, going so far as to favor, in Disraeli's case, extension of the franchise to all adult males.[58]

At the same time, the Chartists' cause occasionally found favor among certain radical members of the industrial middle classes, as was famously the case with the German socialist Friedrich Engels, for whom devil's dust served as a potent metaphor for the dehumanizing situation of the worker. Engels, himself the son of a wealthy textile industrialist, traveled to Northern England in 1844 and published *The Condition of the Working Class in England* (in German) the following year. In a chapter entitled "The Great Towns," Engels used shoddy—termed "Devil's-dust cloth"—to depict the worst aspects of the condition of working people in the country.

In Engels's formulation, devil's dust refers both to a type of working environment and to a product to be worn on the skin; as such, it materializes the extent to which the Industrial Revolution had worsened the plight of the British worker in both mind and body. Engels was struck by the inadequacy of the working class's insulation against the Northern England climate.[59] The workingman, who makes noxious dust, can only wear rags; he who labors in the production of woolens and worsteds is unable to afford either.[60] Clothing, in his account, is a key marker of human adaptation to environment, happiness, and self-definition. And wool, which Engels perceives to be a necessity for the Northern England climate, is available to the middle and upper classes but not the working class.

> The whole clothing of the working-class, even assuming it to be in good condition, is little adapted to the climate. . . . The damp air of England, with its sudden changes of temperature, more calculated than any other to give rise to colds, obliges almost the whole middle-class to wear [wool] flannel next to the skin, about the body, and flannel scarf's [*sic*], and shirts are in almost universal use. . . . Not only is the working-class deprived of this precaution, it is scarcely ever in a position to use a thread of woollen clothing . . . and, if a working-man once buys himself a woollen coat for Sunday, he must get it from one of the "cheap shops" where he finds bad, so-called "Devil's-dust" cloth, manufactured for sale and not for use, and liable to tear or grow threadbare in a fortnight, or he must buy of an old clothes'-dealer a half-worn coat which has seen its best days, and lasts but a few weeks.[61]

The term is similarly used (along with that of "shoddy king") in Karl Marx's sympathetic description of Chartist resistance in the 1850s to the attempts by middle-class industrialists to enlist the aid of the working class in common cause against the landed aristocracy.

> *London*, March 20 [1855]. For several months *The Morning Advertiser* has endeavoured to set up a propaganda society under the name of National and Constitutional Association for the purpose of overthrowing the oligarchic regime. After many preparations, appeals, subscriptions, etc., a public meeting was at last called for last Friday at the London Tavern. It was to be the birthday of the new, much advertised Association. Long before the meeting opened the great hall was crowded with working men, and the self-appointed leaders of the new movement, when they appeared at last, had difficulty in finding room on the platform.[62]

To the would-be leaders' clumsy allusion to the special ability of members of the middle class when it came to governing the country, a leading Chartist (himself a member of the landed aristocracy) indignantly replied that "he had no objection to join at any moment in an endeavour to upset the influence of the Duke of Devonshire, et al., but he would not do so to establish in its stead that of the Duke of Devil's Dust or a Lord of Shoddy," a response that Marx reports was met with cheers and laughter. As Marx added in his own name, to drive home the irony intended by the term "Lord of Shoddy": "Those who refuse to broaden the franchise to cover the whole of the people by adopting the People's Charter are admitting that they wish to replace the old aristocracy by a new one."[63]

The shoddy industry continued to be a focal point in varied political debates for decades to come. As more and more people found themselves clothed in shoddy, those who invoked the language of devil's dust did so in support of diverse causes. Shoddy was not only used as a metaphor for corruption and popular oppression; it was also praised and romanticized, both as a bridge between divisions in social and economic class, and as a symbol of scientific progress and related improvements in the general well-being of society. The status of the "shoddy towns" as, one might say, full-fledged factories of decomposition was something to be celebrated far and wide, according to its promoters. Samuel Jubb, manufacturer and author of *The History of the Shoddy-Trade* (1860), describes "the great rag and shoddy laboratory"[64] as it had developed in the first half of the century:

> We see in the case before us, the principles of economy forcibly and pleasingly illustrated in practice; and materials regarded at one time as almost worthless, converted, by the improving processes of manual labour and machinery, into valuable elements of textile manufactures.[65]

This focus on political economy participated in a more general trend toward the maximization of industrial resources as part of good business practice, a convergence of environmental and business interests.[66] Shoddy was also lauded within a growing discourse of scientific agriculture and social reformers interested in promoting the productive use of waste products and animal, vegetable, and mineral by-products of all kinds. Among these was author and editor Peter Lund Simmonds, whose two editions of *Waste Products and Undeveloped Substances* featured extensive discussions of shoddy's use as agricultural fertilizer, drawing on the emergence of scientific agriculture in the preceding decades.[67]

Throughout this period there was, in short, a more positive overt account, especially associated with the Whiggish celebration of free labor. As we saw in the prologue, historian and businessman Samuel Jubb had it that there was no waste in a shoddy landscape: "Not a single thing belonging to the rag and shoddy system is valueless, or useless; there are no accumulations of mountains of debris to take up room, or disfigure the landscape; all—good, bad, and indifferent—pass on, and are beneficially appropriated."[68] Instead of useless materials, there were only ingredients for productive transformation. And while some social and cultural critics extolled the usefulness of wool waste as a boon to farm productivity, other writers considered its effect on the British landscape in more poetical, though also laudatory, terms. Writing in the 1830s, and in seeming partial imitation of Carlyle, Sir George Head—under titles such as "Rag-Grinders," "A Metaphysical Question," and "Tillage-Muck"—had emphasized the positive, and evocative, class-crossing capacity of shoddy.[69] *Black's Picturesque Guide to Yorkshire*, a guide for tourists first published in 1858, noted that there was one site in Batley potentially worth seeing: "the shoddy mills."[70] Batley offered sites of "cloth transformation" for something that might approximate spiritual reflection—the metamorphosis of wool. Despite or in some cases in light of all the criticism along political lines, shoddy fascinated a subset of poets and middle-class readers.

Those who visited might have seen the working and manufacturing classes all but mingling on the steps of what became known as the "Shoddy Temple" (see prologue). On the steps of this Methodist Church in downtown Batley, it was alleged that the majority of the rag deals were done on Sundays. Shoddy mill workers and owners would all find themselves lined up in wool suits, shoddy or otherwise. For some of the working class, the development of less expensive shoddy fabrics meant that they could for the first time in their lives afford an ostensibly "new" Sunday suit.[71] In this regard, to quote one journalist of the day, "we might almost moralise on the metempsychosis of wool, the transfer of soul from one coat to another."[72]

An interesting and provocative comparison can be made with philosopher Andrew Pickering's contemplation of clothing and formulation thereby of the concept of "mangle" in relation to aspects of Britishness and personal experience with textiles, in his *Mangle of Practice*:

> The parts of the world that I know best are ones where one could not survive for any length of time without responding in a very direct way to such material agency—even in an English summer (never mind a midwestern winter) one would die quite quickly of exposure to the elements in the absence of clothing, buildings, heating, and whatever. Much of everyday life, I would say, has this character of coping with material agency, agency that comes at us from outside the human realm and that cannot be reduced to anything within that realm.[73]

Pickering's "mangle" is of course a theoretical concept, and a potent one at that, applicable to the analysis of both technological and scientific structures. But it is fundamentally and materially rooted precisely in the handling of clothes. A mangle is a mechanical clothing wringer, thus operating at times to aid in the process of cleaning and readying articles for future wear; but a mangle (as instrument) can also serve as what others refer to in the industry as a "mutilator" to be applied to rags no longer credentialed to be used or sold as clothes.

Material Philosophy and the Shredded Self

The diverse ways in which shoddy and devil's dust embodied the political, economic, and moral cleavages that accompanied the emergent industrial technology of mid-nineteenth-century England are epitomized in works by three influential authors: Carlyle, Disraeli, and Marx, at least two of whom were also significant political actors. We shall see that for Carlyle, the stakes come down to the possibility of a reclamation of the self: what I might call a "shredded self." For Disraeli, the dynamic is more explicitly class-based, as opposed to being at the level of the individual. In shoddy, Disraeli sees and invokes the possibility of a (romanticized) bridge between old aristocracy and the workers by means of the industrious treatment of other people's leftovers. Finally, in Marx, we shall see how, quite unexpectedly, devil's dust materializes—epitomizes even—a fundamental tension and uncertainty between waste, on the one hand, and potentiality, on the other (fig. 1.8).

We begin with Thomas Carlyle's "Devil's Dung" (*Teufelsdröckh*), a mysterious and deeply philosophical character who is both the frame and the undoing of the narrative of *Sartor Resartus*. Carlyle is acknowledged today as a seminal figure in the theorization of technology, although he is often (probably wrongly) seen as a first manifestation of what might be called technological pessimism.[74] Carlyle's "Signs of the Times" (briefly discussed above) laid out two kinds of machine: the "outward" and the "inward"; for the inward machine, Leo Marx has observed, Carlyle had not yet discovered fully adequate language. *Sartor Resartus* (The Tailor Repatched [literally: retailored]), on which he was busy working when "Signs of the Times" was published, met that need in ways that would establish Carlyle's literary reputation and exert powerful influence on a wide range of important nineteenth-century writers and thinkers, from Ralph Waldo Emerson to Alfred Lord Tennyson.

Sartor Resartus, which Carlyle once referred to as an "Essay on Metaphors," is framed as an extended review of what turns out to be an imagi-

Fig. 1.8 "The Street-Seller of Crockery-Ware Bartering for Old Clothes," from a daguerreotype by Charles Beard and published in Henry Mayhew's *London Labour and the London Poor* (1851). (Courtesy of Harvard Library.)

nary text on the "philosophy of clothing"[75] by one Diogenes Teufelsdröckh, Professor of "Matters in General" at the University of Weissnichtwo ("I know not where"). *Teufelsdröckh*, which literally means "devil's dung" or "excrement," is the German term for asafoetida (a common emetic), a verbal blend of East (asa = Persian *azā* for "root") and West (foetida = Latin for "stinking"). But "dreck" in German can also be translated as "dirt" or "earth" and is as such not far from "dust."[76] The fictional German book is in two parts: one descriptive/historical (*Werden*), the other speculative/

Fig. 1.9 "Professor Teufelsdröckh of Weissnichtwo." Illustration from Thomas Carlyle's *Sartor Resartus* (1836), made by Edward Sullivan for a late nineteenth-century edition (London: G. Bell & Sons, 1898), p. 21. (Collection of the author.)

philosophical (*Wirken*). In book 1 of the actual tome, the (fictional) editor lays out the task of assembling a biography of the author and the meaning of the text before him. Book 2 presents what we are told is a biography of the author, constructed from several paper bags full of bits and pieces of archival documents. Book 3 brings us back to an exploration of the project in its entirety.

As a work of art, *Sartor Resartus* bears a curious formal resemblance not only to fabric—to textile media, as it were—but also to shoddy itself; in order to find a publisher, Carlyle found it necessary, as he puts it, to "slit [*Sartor Resartus*] up into stripes";[77] the manuscript literally had to "clip itself into pieces." The work finally appeared in eight installments in *Fraser's Magazine*[78] between late 1833 and mid-1834,[79] under the title

Fig. 1.10 "Old Clothes." Illustration from Thomas Carlyle's *Sartor Resartus* (1836), made by Edward Sullivan for a late nineteenth-century edition (London: G. Bell & Sons, 1898), p. 334. (Collection of the author.)

"Sartor Resartus: The Life and Opinions of Herr Teufelsdröckh in Three Books" (figs. 1.9 and 1.10).

Sartor Resartus has a fantastical and creative form, alternately philosophical, literary, and biographical. Although the book is much studied, Carlyle's use of clothing has not been explored as extensively as a key to understanding his mode of critique along with his general attitude toward "industrialization," a term that he coins. Many accounts view his use of clothing as referring to the superfluous, the superficial, the idea of an evacuation of vital interiority. And to some extent, this can be backed up in the text, especially insofar as he uses the terminology of clothing to describe the limits and realities of language itself.[80]

In fact, however, Carlyle's interest in clothing is profound, not just in jest, and is material as well as metaphorical. As stated by his narrator at the very outset: "How, then, comes it, may the reflective mind repeat, that the grand Tissue of all Tissues, the only real *Tissue*, should have been quite overlooked by Science,—the vestural Tissue, namely, of woollen or other cloth; which Man's Soul wears as its outmost wrappage and overall; wherein his whole other Tissues are included and screened, his whole Faculties work, his whole Self lives, moves, and has its being?"[81] Indeed, he goes so far as to call for a "science of clothes."[82] Clothing is not just a metaphor for deception and "anti-spirit" but both exposes and conceals the general conditions of human existence. Clothing for Carlyle is a rich site precisely at the nexus of interior and external experience. Seen in this light, clothes can be examined as the integuments of human culture as a whole, be it industrial or otherwise.[83]

Clothing is at once the medium, the means, and the matter of things, at once the temporary "thatching" of the fixed patterns by which "victims" of circumstance "stand lashed together uniform in dress and movement."[84] Clothes seem to be emblems, for Carlyle, when they are whole—cloaks, aprons, signifiers of a specific class or station. But clothes in process, as material, can never be pure emblem insofar as they are also always in transition. Here it is the "grind"—the process of rubbing things down (whether through human wear or machine grinding) that gets us to something tangible, material, the essence of things, in a sense. Clothes are thick at their core; a result of their formation through the building up and tearing apart of rags gathered from art and nature.

In book 1, chapter 8 ("The World Out of Clothes"), Carlyle provides an extract from Teufelsdröckh's "speculative part" on the *Wirken* (effects, working) as distinguished from the *Werdern* (origins, becoming) of clothes. Among other things, he bemoans the plight of the human being, not equipped to exist without clothing intentionally, actively, and perpetually remade. The question then becomes, what does the age of machinery do to clothing, and what can looking at clothing tell us about our relationship to machinery?

> It was in some such mood, when wearied and foredone with these high speculations that I first came upon the question of Clothes. Strange enough, it strikes me, is this same fact of there being Tailors and Tailored.

> The Horse I ride has his own fell: strip him of the girths and flaps and extraneous tags I have fastened round him, and the noble creature is his own sempster and weaver and spinner: nay his own bootmaker, jeweller, and man-milliner; he bounds free through the valleys, with a perennial rainproof court-suit on his body; wherein warmth and easiness of fit have reached perfection; nay, the graces also have been considered, and frills and fringes, with gay variety of colour, featly appended, and ever in the right place, are not wanting. While I—Good Heaven!—have thatched myself over with the dead fleeces of sheep, the bark of vegetables, the entrails of worms, the hides of oxen or seals, the felt of furred beasts; and walk abroad a moving Rag-screen, overheaped with shreds and tatters raked from the Charnel-house of Nature, where they would have rotted, to rot on me more slowly! Day after day, I must thatch myself anew; day after day, this despicable thatch must lose some film of its thickness; some film of it, frayed away by tear and wear, must be brushed off into the Ashpit, into the Laystall; till by degrees the whole has been brushed thither, and I, the dust-making, patent Rag-grinder, get new material to grind down.

Writing in the early 1830s, deeply affected by his earlier excursions into German literature and theoretical criticism, Carlyle sees in the industrial transformation of England a Mephistophelean complement, projected on the "screen" of clothes, to the spiritual pyrotechnics of speculative philosophy. In this unique and uniquely influential exercise, the ambiguous figure of "Devil's Dung" (whose first name is Diogenes, recalling both the ancient cynic and, taken more literally, the "gift of the gods") provides a frame onto which the author hangs his own projective screen of dust. Still strongly identified with the Reform movement (he would publish a famous essay supporting the Chartist movement in 1838), Carlyle had not yet embraced the anti-democratic tide with which his later praise of "heroes" and "hero-worship" is generally associated.[85] And yet the idiosyncratic voice of Devil's Dung—alternatively irritating and alluring, emerging as it does ventriloquistically from the mouth of Carlyle's fictional editor—not only illustrates like few others the ineluctable duality of language, but also presages Carlyle's later championing of the individual and the exceptional over the mass and common, of the one nation over the other, in contrast to the next thinker to whom we turn.

"Devilsdust" and the Protean Nature of Disraeli's Shoddy

In 1844, the year before Friedrich Engels's *The Condition of the Working Class in England* appeared, Benjamin Disraeli published his novel *Sybil, or The Two Nations*, with the explicit intention of arousing indignation over the squalor in which the English working classes lived and labored. Disraeli, whose father had converted to Christianity when Disraeli was twelve, had served as a member of Parliament since 1837, where he had emerged as a vocal opponent of the Corn Laws and an early sympathizer with the Chartist movement, which sought universal male suffrage and other parliamentary reforms. He was not yet the famous statesman he would become as a forceful prime minister and leader of the Conservative Party; but he already evinced interest in effecting an alliance between the landed aristocracy and the industrial working class in contrast to the rigid liberalism of the urban industrialists—that is, to weave together a single nation (to adopt the terms of the novel's subtitle).

Sybil is literally about the marriage of Disraeli's two favored classes: Sybil Gerard is a Chartist, while her lover and later husband is Charles Egremont, an aristocrat moved by the plight of the working poor. Their romance is carried out against the backdrop of labor agitation and increasing radicalism and violence, to which some of the Chartists succumb, while holding out the promise of a more peaceful, meliorist solution, as embodied in the union of this couple. But the more interesting character for our purposes does not figure in the "plot" at all, though he weaves among its many chapters as a kind of cipher. By the time of Disraeli's writing, amidst political upheavals from the Corn Law Riots to the debates about Chartism and the Factory Acts, devil's dust was, as we have seen, both the synonym for and by-product of shoddy. An anonymous figure, "the nameless one"—other characters call him "Devilsdust"—emerges from the diseased aura of the rags among which he lives, and from which it seems he has sprung, rags apparently polluted with the filth of poverty, of what Disraeli would develop as "the two nations"—a material incarnation of rags as symbol of both despair and possibility.

Devilsdust is an orphan of unknown parentage (there is some suggestion of aristocratic lineage on his father's side), who (almost) miraculously survives the deprivations of his youth, themselves emblematic of the sufferings of many. Devil's dust, as used by Ferrand and Engels, represented—materialized even—the ills and unpleasantnesses of the factory system along with working-class decrepitude more generally. Disraeli's characterization of this mysterious person, by way of contrast, points to a more redemptive role for shoddy. Disraeli's development of the Devilsdust character suggests the protean potentiality inherent in rags, which also makes them vectors for disease, disgust, and everything in between. Devilsdust is born in squalor, "a nameless one," and at the same time "a vital principle" who inexplicably survives into robust adulthood. The character is introduced, in all his mysterious and metaphorical glory, in this passage, quoted in full for effect:

> "Well, Devilsdust, how are you?"
>
> This was the familiar appellation of a young gentleman, who really had no other, baptismal or patrimonial. About a fortnight after his mother had introduced him into the world, she returned to her factory and put her infant out to nurse, that is to say, paid threepence a week to an old woman who takes charge of these new-born babes for the day, and gives them back at night to their mothers as they hurriedly return from the scene of their labour to the dungeon or the den, which is still by courtesy called "home." The expense is not great: laudanum and treacle, administered in the shape of some popular elixir, affords these innocents a brief taste of the sweets of existence, and keeping them quiet, prepares them for the silence of their impending grave. Infanticide is practised as extensively and as legally in England, as it is on the banks of the Ganges; a circumstance which apparently has not yet engaged the attention of the Society for the Propagation of the Gospel in Foreign Parts. But the vital principle is an impulse from an immortal artist, and sometimes baffles, even in its tenderest phases, the machinations of society for its extinction. There are infants that will defy even starvation and poison, unnatural mothers and demon nurses. Such was the nameless one of whom we speak. We cannot say he thrived; but he would not die. So at two years of age, his mother being lost sight of, and the weekly payment having ceased, he was sent out in the street to "play," in order to be run over. Even this expedient failed. The youngest and the feeblest of the band of victims, Juggernaut spared him to Moloch.

> All his companions were disposed of. Three months' "play" in the streets got rid of this tender company,—shoeless, half-naked, and uncombed,—whose age varied from two to five years. Some were crushed, some were lost, some caught cold and fevers, crept back to their garret or their cellars, were dosed with Godfrey's cordial, and died in peace. The nameless one would not disappear. He always got out of the way of the carts and horses, and never lost his own. They gave him no food: he foraged for himself, and shared with the dogs the garbage of the streets. But still he lived; stunted and pale, he defied even the fatal fever which was the only habitant of his cellar that never quitted it. And slumbering at night on a bed of mouldering straw, his only protection against the plashy surface of his den, with a dungheap at his head and a cesspool at his feet, he still clung to the only roof which shielded him from the tempest.
>
> At length when the nameless one had completed his fifth year, the pest which never quitted the nest of cellars of which he was a citizen, raged in the quarter with such intensity, that the extinction of its swarming population was menaced. The haunt of this child was peculiarly visited. All the children gradually sickened except himself; and one night when he returned home he found the old woman herself dead, and surrounded only by corpses. The child before this had slept on the same bed of straw with a corpse, but then there were also breathing beings for his companions. A night passed only with corpses seemed to him in itself a kind of death. He stole out of the cellar, quitted the quarter of pestilence, and after much wandering laid down near the door of a factory. Fortune had guided him. Soon after break of day, he was woke by the sound of the factory bell, and found assembled a crowd of men, women, and children. The door opened, they entered, the child accompanied them. The roll was called; his unauthorized appearance noticed; he was questioned; his acuteness excited attention. A child was wanted in the Wadding Hole, a place for the manufacture of waste and damaged cotton, the refuse of the mills, which is here worked up into counterpanes and coverlids.[86]

Devilsdust grows up amidst the squalor of rags, ragpickers, and the decaying remains of both the dead and dying. And yet, from these rags, from devil's dust and worse, much good can come:

> The nameless one was preferred to the vacant post, received even a salary, more than that, a name; for as he had none, he was christened on the spot—*DEVILSDUST.*
>
> Devilsdust had entered life so early that at seventeen he combined the experience of manhood with the divine energy of youth. He was a first-rate workman and received high wages; he had availed himself of the advantages of the factory school; *he soon learnt to read and write with facility, and at the moment of our history, was the leading spirit of the Shoddy-Court Literary and Scientific Institute* (emphasis added).[87]

Devilsdust finds a name, and then a livelihood, in what might have been seen as his death sentence, eventually emerging, improbably enough, a successful capitalist, unlike most of his less fortunate (or protean) proletarian companions.

What is one to make of the presence of such a figure, who neither drives the plot nor participates significantly in its unfolding, other than as an indication of disturbing possibilities, superficially suppressed, inherent in a system of exchange, materialized in shoddy itself, in which nothing is permanent or stable. Tellingly in this regard, the crucial speech calling for the creation of a single "nation" is put into the very ambiguous mouth of Stephen Morley. The only reason can be that Disraeli is much more concerned to point to the dangers implicit in the two nations than to offer any concrete "one nation" solution. So Disraeli can wholeheartedly endorse the socialist critique of society—and Morley (as yet unnamed—perhaps because Disraeli sees there was a problem in reconciling the two nations speech with Morley's later development) is allowed to chill his readers' blood with the threats inherent in a class-divided society. But Disraeli cannot begin to attribute any legitimacy to Morley's socialist solution, and so Morley is transformed from hero to villain. The only working-class characters who ultimately thrive are Devilsdust and his companion Dandy Mick Radley, both of whom enter enthusiastically into the rioting, but then abandon their "ideals" in order to become capitalists, and whose descendants, as the author/narrator promises, will eventually be incorporated into the aristocracy. This promise reflects a typical meritocratic myth—that the most talented members of the working class can rise out of it. Of course, Disraeli's belief in the hereditary principle (and his own [fictionally] aristocratic roots)[88] is compatible with this position, as it is with the "unknown" character of Devilsdust's own (paternal) origins.

Shoddy as Paradox and Marx's "Excrements of Consumption"

Such disturbing possibilities become famously—even notoriously—explicit in the writings of Marx and Engels, to whose own *Condition of the Working Class in England* we have already had occasion to refer. As the scion of a family of German textile industrialists, Engels was quite familiar with the "dust" that both covered the shop floor and was later sent to the manure heap, while at the same time serving as a shorthand synonym for shoddy itself as both a product and a system of production. For Marx as well, who had himself made use of the term in the mid-1850s, devil's dust figures as an important bearer of significant economic and political meaning. Marx, who first met Engels in 1842 when the latter was en route through Prussia to Manchester, and who would later collaborate with him on *The Communist Manifesto* (1848), singled out for criticism not only the rag itself, but also the industrial and social context of their processing and production. Rags appear in all the volumes of *Capital* starting with the first in 1867. Marx discusses devil's dust in the important chapter 8 of *Capital*, volume 1, "Constant Capital and Variable Capital."

> Suppose that in spinning cotton, the waste for every 115 lbs. used amounts to 15 lbs., which is converted, not into yarn, but into "devil's dust." Now, although this 15 lbs. of cotton never becomes a constituent element of the yarn, yet assuming this amount of waste to be normal and inevitable under average conditions of spinning, its value is just as surely transferred to the value of the yarn, as is the value of the 100 lbs. that form the substance of the yarn. The use-value of 15 lbs. of cotton must vanish into dust, before 100 lbs. of yarn can be made. The destruction of this cotton is therefore a necessary condition in the production of the yarn. And because it is a

> necessary condition, and for no other reason, the value of that cotton is transferred to the product.[89]

Devil's dust here stands not for the product (i.e., shoddy) but the refuse that is its by-product, and yet Marx is impatient to insist that its full "value" (as distinguished from its matter) is transferred over to the usable yarn, whose own value, in accordance with the theories of two classical economists Adam Smith and David Ricardo that Marx here appropriates, is entirely a function of the value of the labor, whether direct or indirect, necessary for its production. While the matter of the dust may end up on the waste heap, in other words, its value remains in circulation, no less than does the finished fabric itself.[90]

The paradoxical character of devil's dust emerges with even greater force in volume 3 of *Capital* (a compilation of Marx's writings composed between 1863 and 1883), in which Marx (heavily edited by Engels) claims, in quick succession, that such dust drops out of the productive cycle (as what he calls an "excrement of consumption") and that it remains a vital part of it. In part 1, chapter 5, section 1, they write that just as "continual improvements" in the economy of production presuppose the cooperation of laborers, so too with "the second big source" of such economy: namely, "the reconversion of the excretions of production, the so-called waste, into new elements of production"—that is,

> to the processes by which this so-called excretion is thrown back into the cycle of production and, consequently, consumption, whether productive or individual. This line of savings, which we shall later examine more closely, is likewise the result of large-scale social labour. It is the attendant abundance of this waste which renders it available again for commerce and thereby turns it into new elements of production. It is only as waste of combined production, therefore, of large-scale production, that it becomes important to the production process and remains a bearer of exchange-value. This waste, aside from the services which it performs as new elements of production, reduces the cost of the raw material to the extent to which it is again saleable, for this cost always includes the normal waste, namely the quantity ordinarily lost in processing. The reduction of the cost of this portion of constant capital increases *pro tanto* the rate

> of profit, assuming the magnitude of the variable capital and the rate of surplus-value to be given.[91]

To the extent that waste, rather than merely being minimized, can itself enter directly into the cycle of production, it increases the rate of surplus value and hence the profit of the capitalist.[92]

There is the topic of "the sewage question," which is addressed in various articles of literature, specifically thinking about the value of waste (human waste in particular, but the idea is extendable to textile waste) in relation to issues surrounding public health and disease prevention. Sewage (aka night soil) was valued insofar as it could be used for fertilizer, but specific knowledge of its health risks increased. This duality—of both the value and dangers inherent in waste—extended to rags, as in the case of Devilsdust from Disraeli, who emerged from a sickening environment of polluted rags and yet was able to generate value from precisely that matter. By the 1870s, with Disraeli between his two stints as prime minister, he was calling for an explicit focus on sanitary legislation.[93] Meanwhile, skepticism of decades of promotion of the idea of human waste as being commercially valuable grew more widespread, moving beyond the confines of efforts to establish profitable agricultural "sewage farms" to towns and cities that increasingly had to pay for the treatment and disposal of waste.

And yet whereas, in general, Marx is extremely interested in reuse—its necessity and yet often absence in capitalism—shoddy seems to get under his skin. Marx proceeds to contrast "rags," which he here calls an "excrement of consumption," comparable to the excretions of the body deposited as night soil, with "excrements of production" like "iron turnings," which remain part of the production process.

> With the advance of capitalist production the utilisation of the excrements of production and consumption is extended. We mean by the former the refuse of industry and agriculture, and by the latter either the excrements, such as issue from the natural circulation of matter in the human body, or the form in which objects of consumption are left after being used. Excrements of production, for instance in chemical industries, are such by-products as are wasted in production on a smaller scale; iron filings collected in the manufacture of machinery and carried back into the pro-

> duction of iron as raw material, etc. Excrements of consumption are the natural discharges of human beings, remains of clothing in the form of rags, etc. The excrements of consumption have the most value for agriculture. So far as their utilisation is concerned, the capitalist mode of production wastes them in enormous quantities. In London, for instance, they find no better use for the excrements of four and a half million human beings than to contaminate the Thames with it at heavy expense.[94]

Only a few paragraphs later, however, the author(s) correct this misleading and technically inaccurate description of rag waste:

> The wool industry was carried on more intelligently than the preparation of flax. The same report states on page 107 that it was formerly the custom to veto the preparation of waste wool and woolen rags for renewed use, but this prejudice has been entirely dropped so far as the shoddy trade is concerned, which has become an important branch of the wool district of Yorkshire. It is doubtless expected that the trade with cotton waste will soon occupy the same rank as a line of business meeting a long felt want. Thirty years previous to 1863, woolen rags, that is to say pieces of all-wool cloth, etc., were worth on an average about 4 p.st. 4 sh. per ton. But a few years before 1863 they had become worth as much as 44 p.st. per ton. And the demand for them had risen to such an extent that mixed stuffs of wool and cotton were also used, means having been found to destroy the cotton without injuring the wool. And thousands of laborers were employed in 1863 in the manufacture of shoddy, and the consumer benefited thereby, being enabled to buy cloth of good quality at very reasonable prices. The shoddy so rejuvenated constituted in 1862 as much as one-third of the entire consumption of wool in English industry, according to the factory report of October, 1862, page 81. The truth about the "benefit" for the "consumer" is that his shoddy clothes wear out in one-third of the time which good woolen clothes used to last, and become threadbare in one-sixth of this time.

One is tempted to suspect that the earlier passage is from Marx's own hand, while the latter was added by Engels, whose personal familiarity with the shoddy industry here intrudes upon Marx's own classification of rag refuse as an "excrement of consumption." Erroneous or not, that classification may reflect a deeper insight on Marx's part (not necessarily shared by Engels) as to shoddy's peculiar status as a stand-in for the "consumed" worker himself. Even Engel's "correction," however, does not save shoddy from the author(s)' ultimate suggestion that "economy" is here purchased at too great a price, at least from the standpoint of the "consumer," inasmuch as the ultimate wearer of the manufactured clothing and the worker who produces it are here one and the same. In any case, by characterizing shoddy as both an "excrement" of consumption and one of production, volume 3 of *Capital* draws attention to the interchangeability of production and consumption especially as it bears upon the human body (whose waste is explicitly compared to that of rags).

As for that deeper insight: in *Capital*, volume 1, devil's dust (here understood as what volume 3 would call an "excrement of consumption") is indeed a stand-in for the wasted and consumed worker, whose labor power, and accompanying value, is transferred to, and remains alive, in the product of his labor. In other words, surplus value is to the worker, wasted in his decrepitude, as shoddy blankets are to devil's dust. Indeed, it is not the individual worker, but the workers as a social mass, with which devil's dust is most accurately identified.[95]

This implicit identification of devil's dust and the workers themselves may partly explain Marx's curious choice of the term "lumpenproletariat"—a word that he coined—to name the refuse of society who cannot be productively absorbed into the class struggle. While some scholars have used Marx's term "devil's dust" to refer to emptied-out value itself,[96] it may be more helpful to link it, both positively and negatively, with "Lumpen." The German *Lumpen* literally means "rags" or "shreds," and figuratively "rascal"; *in Lumpen sein* means "to be in rags and tatters." As with the "shoddy-hole," in which, as Marx says, "rags are pulled to pieces" and children forced to labor "ceaseless[ly]," devil's dust represents both the promise of revolution and its hellish opposite. In *The Eighteenth Brumaire of Louis Napoleon*, published in 1852, Marx famously defines the lumpenproletariat as follows:

> Alongside decayed roués with dubious means of subsistence and of dubi-

> ous origin, alongside ruined and adventurous offshoots of the bourgeoisie, were vagabonds, discharged soldiers, discharged jailbirds, escaped galley slaves, swindlers, mountebanks, *lazzaroni*, pickpockets, tricksters, gamblers, *maquereaux* [pimps], brothel keepers, porters, *literati*, organ grinders, ragpickers, knife grinders, tinkers, beggars—in short, the whole indefinite, disintegrated mass, thrown hither and thither, which the French call *la bohème*.[97]

Marx proceeds to link the lumpenproletariat with the rentier (or landlord) class, neither of which in his account plays a productive social role. He makes a similar point in *The German Ideology*, coauthored with Engels,[98] in which "lumpenproletariat" appears for the first time, in a section partly devoted to describing the foiled efforts of the Chartists between 1839 and 1844 to empower English workers.[99]

For Marx, in short, the situation of devil's dust comes to represent humanity itself both as a proletarian "universal class" and as its "lumpen" inversion (fig. 1.11). Far more than in his reflections on other industries, Marx seems practically revolted by the rag business, describing rag sorters in dehumanizing terms in his chapter on "Machinery and Large-Scale Industry" in a section on "Modern Manufacture." As Marx there writes:

> One of the most shameful, the most dirty, and the worst paid kinds of labour, and one on which women and young girls are by preference employed, is the sorting of rags. It is well known that Great Britain, apart from its own immense store of rags, is the emporium for the rag trade of the whole world. They flow in from Japan, from the most remote States of South America, and from the Canary Islands. But the chief sources of their supply are Germany, France, Russia, Italy, Egypt, Turkey, Belgium, and Holland. They are used for manure, for making bed-flocks, for shoddy, and they serve as the raw material of paper. The rag-sorters are the medium for the spread of small-pox and other infectious diseases, and they themselves are the first victims. . . . They become rough, foul-mouthed boys, before Nature has taught them that they are women. Clothed in a few dirty rags, the legs naked far above the knees, hair and face besmeared with dirt, they learn to treat all feelings of decency and of shame with contempt.[100]

Fig. 1.11 "The whole indefinite, disintegrated mass" (text from Marx's *The Eighteenth Brumaire*). Photo by Hanna Rose Shell.

ACT II

Textile Skin

> Oh my country!
> Could it be you did intend,
> Wretches draped in shameful shoddy
> To the battlefield to send?
> Shoddy ripping, shoddy bursting,
> Shoddy rotting in a day.
>
> "SONG OF THE SHODDY," *VANITY FAIR* (1861)

On the eve of the firing on Fort Sumter on April 12, 1861, as men such as Oliver Wendell Holmes Jr. were preparing themselves for war, so too was the industrial infrastructure. The same manufacturers, and indeed the same fiber blends, that had been outfitting the human bodies of laboring slaves in "negro cloth," as it was generally called, were now being readied for soldiers fighting for the cause of abolition.

Holmes would have shipped out carrying a French musket. This 1861 *carte de visite* (an albumen print posted on paper card of approximately three-by-four inches, slightly smaller than modern-day snapshot size) shows him posing in his dress uniform prior to departing for the front (fig. 2.1). Proud and determined in facial expression, he cradles in the crook of his elbow as it rests upon his knee not a musket but a sword and scabbard. His elegant dress uniform displays a row of brass buttons down the front, as well as plush brown epaulets and gold piping—signifying rank—both here and down the sides of his trousers. It would prove to be a highly idealized image viewed in the sober light of the conditions he would soon face

Fig. 2.1 Oliver Wendell Holmes Jr. in uniform, *carte de visite*, 1861, made in Boston by Silsbee, Case & Co. photographers. (Courtesy of Boston Public Library.)

and a stark contrast with the actual uniforms most soldiers would take to the battlefield. Holmes's dark blue woolen cap, or kepi, bears a bugle insignia marked "20," the number assigned to his regiment.

Such photographs of soldiers in dress uniform were common ways of showing preparedness and readiness, not just of the individual, but also of the army as a whole. Photographic *cartes de visite* had become widely popular in 1859 in France, a few years after the technique's invention in 1854.[1] They arrived in the United States just in time for the start of the Civil War and were quickly adopted as a convenient way for family members and friends to keep fond memories alive during the military deployment of their loved ones.

Along with photography, other kinds of supplies and accompanying technologies were also being prepared. Holmes shipped out as a lieutenant for the Massachusetts Twentieth Infantry in September 1861, just as the word *shoddy* in a newly metaphorical sense was entering common discourse.[2] And as we shall see in this act, over the course of the Civil War, shoddy became a simultaneously concrete and abstract target. An enduring subject of government contracting scandals, shoddy at the same time came to materialize both real and metaphorical linkages between slaves and soldiers, between abolitionism and the pull of capital. As part of this process, *shoddy* as a term experienced "adjectification" and personification. It came to fill a "semantic void," a word for which all the conditions had been met but had not yet appeared, as well as providing a verbal medium for the naming of and critique of (what we would call today) the nouveau riche.[3]

Graphical and poetic images of shoddy in this period, including photographic depictions of shoddy blankets (which often served as makeshift shrouds on Civil War battlefields), illustrate shoddy's function as material, as medium, and as mode of representation. Shoddy is a talismanic "textile skin," a term introduced here to evoke the uniquely interstitial nature of shoddy. The textile skin, as a worn or worn-out stretch of cloth, is—as I develop this concept—a layer applied to the body that both protects it and exhibits it, a border zone between the naked individual and her world. Once discarded, shredded, and rendered as cloth, it becomes a bearer of its former owners and an imbrication of them together into a collective mass.

Discourses of shoddy and of photography converge in this period. The following analysis of shoddy in the Civil War culminates in a close study of a famous series of documents, photographic "skins," as Oliver Wendell Holmes Sr. might have called them, made of corpses on the battlefield at Gettysburg. Indeed, the association between the photograph as a kind of eternal fiber-and-emulsion material potentially replacing the more transient skins of living nonhuman beings had been raised by Holmes in 1859 in a popular essay wherein the photographic negative is compared to the bark of a tree, or the pelt of an animal, in the context of "forms, effigies, membranes" and "evanescent films." The textiles depicted in the Alexander Gardner photographs materialize a series of otherwise hidden contexts. They already bear the traces of the context in which these bodies came to cease living; and they instantiate the conditions that led to the war—and its contradictions. At the same time, these sartorial textile skins form an element of another skin, the fibrous and emulsion-coated photographic print itself.[4] We shall see how the printed photograph emerges as its own fiber art; within their emulsion sinews, the shoddy uniforms will remain etched by light.

The Wear of War

The sudden need for uniforms overwhelmed the ability of Northern cloth and clothing producers to provide them quickly, while also furnishing lucrative opportunities for personal enrichment. Abraham Lincoln's first declaration of war in April 1861 triggered an almost immediate and overwhelming demand for uniforms and related paraphernalia.[5] Manufacturers such as Brooks Brothers resorted to the use of inferior shoddy fabrics that performed poorly and resulted in a series of inquiries and related scandals that would reverberate widely, eventually investing the term *shoddy* with a new and disturbing set of meanings it had not previously explicitly harbored.

Shoddy had originally arrived in the United States as a saleable manufacture in the form of bolts of cloth.[6] The lore surrounding the emergence of shoddy on the British side connects this to the development of markets in the American South. Indeed, Benjamin Law, who has often been credited as shoddy's "inventor," seems to have almost immediately perceived a market for his goods in the United States, first launching a trade mission in the early years of the nineteenth century. Slave owners would need to provide blankets to help their slaves sleep at night, but presumably wouldn't want to pay a great deal of money for those blankets. Law figured he could sell shoddy blankets to slave traders at prices more attractive than any they'd seen before for woolen products, and so packed up a large consignment of blankets and cloth and set sail with his son, John, seventeen years old at the time. They disembarked in New York, transported the consignment down to the Southern states, and sold out the whole lot. After they returned to Batley, flush with success, they were invited to give a talk about their adventures at the town's newly established Sunday school, in which Law excitedly described not only the overseas markets but also more of the details of his machinery.[7] Law's son John returned to the United States, heading to Louisiana, where he was never heard from again (presumed to have died from yellow fever), and Law himself remained in West Yorkshire until his death in 1837. Meanwhile, others in town rapidly filled their places and their own coffers, and the shoddy-making industry grew along with the towns of Batley, Dewsbury, and the neighboring villages. Shoddy exports—of both ground material and finished shoddy cloth—took off; lower tariffs on what was often declared as "waste" encouraged this further.

If shoddy first left as finished textile product and then as fibrous material ingredient, it soon thereafter arrived in the United States as process and machinery. And whereas the woven form of the commodity had flowed readily to the slaveholding South, the "raw" material, as well as the machinery and method, found a ready home in the Northern states. Acts of concealment and clandestineness facilitated the technology's export and ultimately the infrastructure's cross-Atlantic diaspora. It is said that the son of one of the blacksmiths who had worked with Law and George Parr in developing the original mechanical rag picker smuggled a prototype "devil" out of the country in defiance of laws prohibiting the export of British machinery or blueprints. He shipped it out through Liverpool, sending it to a contact in New York, having falsely declared it a rice thresher. Although he had intended to follow the machine, he evidently was caught and arrested by British customs, but only after the shrouded rag picker had already shipped. The devil reached American shores a few weeks later. Within a decade, shoddy manufacturers flourished in New York, Pennsylvania, Massachusetts, and Rhode Island, with Rhode Island being the largest provider to plantations of "union" fabric (shoddy cloth mixed with cotton) for the production of work clothing and blankets for slaves (fig. 2.2).[8]

One can read much from even a small swatch of such "negro cloth," in which myriad relations are entwined. In the cloth sample that appears in figure 2.2, for example, which contains a blend of cotton and wool, political tensions both within the United States and on either side of the Atlantic Ocean, as they would have existed in the 1830s, are woven together. The cotton is new, "firsthand" as it were, produced in the South for spinning in New England.[9] The wool, by contrast, is a blend of "new,"

Fig. 2.2 Sample of "slave cloth" made of shoddy cloth included in 1845 correspondence between the Hazard family and slaveholding customers. (Courtesy of Harvard Library.)

"leftover" (clippings), and "used." The blend of wool leans heavily to the "used," as compared to the worsteds being produced at the time for regular (non-slave) wear. As for the "used" bits of wool ground up to make the "wool" part of the thread, these would have been largely drawn from (free) white people's clothes from the Northern states.[10] Much of the raw wool in the original garment, by contrast, would have been sourced outside the country.

The Hazard family of Rhode Island, for example, whose company produced the swatch shown in figure 2.2, built an enormous textile empire by selling negro cloth (at times referred to as "slave cloth" or "plantation cloth") to Southern plantation owners, concurrent with the rapid and very concentrated growth of woolen and worsted manufacture in Rhode Island, eastern Massachusetts, and southern New Hampshire in the 1840s and '50s.[11] Several of the Hazard brothers worked together selling and taking orders for cloth and finished garments for plantations throughout Alabama, Mississippi, and Louisiana. Orders sometimes contained slave names and rough measurements. At other times, bolts were requested, because clothes would be cut and sewn by the enslaved in the Southern states.[12] Shoddy thus entered the American popular vocabulary already haunted by disquieting links with both contraband trade and with the industry of slavery, long before its emergence as a metaphor for Civil War–era corruption and battlefield suffering.

Oliver Wendell Holmes Jr. left for the battlefield imbued with the idealism and passion of youth and the refinement of an upper-class New England upbringing, the child of an ardent abolitionist and a public intellectual. As he set out with what would go on to be called "the Harvard regiment," alongside the sons of a number of New England textile mill owners, at least two shocks were to confront him: both the carnage of the battlefield and concerns, already brewing, about the deep corruption in government contracting for the provision of soldiers' blankets and uniforms (fig. 2.3).

A letter to a Washington official dated February 3, 1861, from New York politician and prison warden Isaac Comstock, even before Lincoln's call to arms in April of that year, documented the problem of low-quality blanket manufacture and combined it with the general issue of both shoddy and government fraud.[13] However, the scale was now infinitely vaster. "As your committee is searching after frauds by government contracts, I feel it my duty as a citizen to render what aid I can in so important a matter. . . . These blankets contain 15 percent wool, the balance entirely made from shoddy."[14] The general public, too, started hearing about problems with uniforms in the field. Two months into the war, a July 15 brief on the condition of troops based at Newport News and Fort Monroe, both in Virginia, was featured in the July 20 *New York Times*, after having also appeared in local newspapers throughout the state, reporting, "Several of the regiments located in this vicinity, and at Newport's News, have suffered severely from the lack of shoes; clothing, camp equipage, and various articles of prime necessity." The report continued:

> Shoddy uniforms, inferior shirts, and shoes of the frailest and shabbiest make, have literally fallen from the limbs of the soldiers of the Empire State, reducing our forces to ragged regiments and shoeless brigades. Some time since it was stated that the New-York troops, who were the victims to these disgraceful swindles, were to receive other uniforms of a better kind. Thus far, however, only the echo of a promise has reached us.[15]

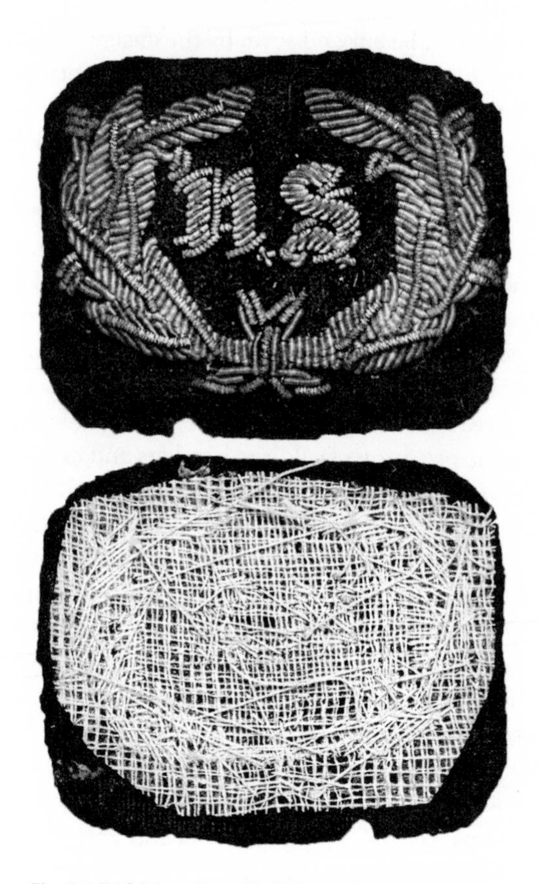

Fig. 2.3 "H.S." American Civil War uniform patch, worn by Oliver Wendell Holmes, front and back. (Courtesy of Harvard Library.)

Less than three months after the official start of the war in April 1861, Holmes was still in Massachusetts. Meanwhile, the Peace Dale Manufacturing Company was in the midst of transitioning in part from the production and marketing of negro cloth to shoddy blankets for military use.[16] The market for the former had dried up with the buildup to war, because the Confederate states increasingly did such business with Britain.[17] In addition, the ethics of making clothing for slaves had become increasingly morally and politically problematic. Participants in the so-called "clean hands" movement, for example, avoided the use of cotton clothing that did, or might, have its origins in the Southern slave plantations. "Cotton thread holds the Union together," Ralph Waldo Emerson had written in 1846, a decade after his enthusiastic publication and endorsement of *Sartor Resartus* in its first printing as a complete manuscript. And yet, such boycotters were certainly content to do business with Northern wool textile manufacturers. Indeed, many of the selfsame manufacturers who claimed moral outrage at slavery were, at the same time, doing brisk business producing and selling shoddy wool blends to plantation owners (in this regard, Roland Hazard proved an exception among the Peace Dale brothers).

New England productive technologies geared toward the manufacture of slave cloth proved less than ideal for the outfitting of soldiers, however. The textile industry was at the time largely centered in New England, where a fine network of rivers supplied ample power for the newly mechanized mills that had sprouted up throughout the region during the previous two decades, replacing the domestically based labor that had prevailed early in the century. During the decades leading up to the Civil War, as the trade with the plantations had decreased, developments in industrial infrastructure, machine technology, and supply chains enabled the rapid growth of the industry. At the same time, the so-called "Tariff of Abominations" of 1828[18] doubled virtually overnight the cost of importing foreign "raw" or "new wool," incentivizing the salvage and reuse of secondhand materials throughout the country, though with a particularly strong impact on the South. Since domestic wool production remained limited, New England textile manufacturers instead turned to the regional supply of rags, often taking the leftovers from the paper industry, as well as the growing supply of low-quality imported rags, which (under the title of "waste") entered the country under lower tariff rates than new wool.

During this period as well, new methods of "carbonizing," which had originated in Europe, were introduced to America. Carbonizing allowed rags and old clothing made of both wool and other vegetable-based fibers, such as cotton linen or jute, to be broken down, making it possible to separate out the degraded wool fibers for shredding and reuse in what was known as "shoddy extract." This opened up, in turn, new possibilities for shredding and reusing the increasingly abundant "union cloth" (fabric spun from mixed materials such as wool and linen or cotton), often used for underwear. Meanwhile, single-swift rag pickers gradually gave way to two-swift and three-swift models. Among the latter was the still ubiquitous "garnett," which combined shredding and carding into a single mechanical process and received a US patent in 1856.[19] With the increasing volume of rags being processed, grading also became more and more refined; this made possible, in turn, more efficient shredding and the tailoring of machinery to suit particular grades of fabric, and it also facilitated the transformation of mungo (shorter-fibered ground rags, generally produced from old felt or reused worsteds, as introduced in the previous act) into "new," or rather "renaissance," as it would come to be called, thread (fig. 2.3).[20]

While the new technologies made the mechanical spinning of shoddy thread cheaper and easier than before, it also encouraged the production of a lower grade of fabric, especially in the US, where new wool was more difficult to come by and the large market for slave goods was especially alluring. The slave cloth industry that flourished in the North had pioneered certain shoddy-based textile "blends," whose purchasers explicitly wanted inferior materials (both to keep costs down and to keep their wearers "in their place").[21] Uniforms and blankets (made partially of shoddy) in the European context, for use either in continental Europe (as for example, in the Crimean War) or by the Confederates during the Civil War, tended to be all wool, part of it reused wool. In the US, by way of contrast, fabrics were more likely to be a blend of (shoddy) wool and linens and cotton materials, what was known in the industry as "linsey-woolsey," or "druggets." The percentage of shoddy as compared to "new" wool used in these goods was much higher than in the British-made uniforms and even in US-made uniforms produced for the Mexican-American War. Civil War blankets, for example (now highly collectible), were exclusively shoddy wool and cotton blends, known as union cloth.

Whereas a whole range of materials was needed for military deployment, wool was in particular demand, greatly exceeding supply.[22] At the same time, the declaration of war had cut off the Northern textile manufacturers from their customer bases in the Southern plantation states. And with cotton especially scarce with Southern supplies cut off, wool became only more desirable. Although the "heavy" wool manufacturers that had been supplying so-called negro cloth and sometimes cotton-linsey (a wool/cotton-blend/union cloth) now had more capacity, it was not nearly enough. According to later accounts (by the Federal Trade Commission in 1918 and congressional records from the 1904 shoddy hearings), at least thirty exclusively shoddy-making establishments operated in 1859, just prior to the outbreak of the Civil War—that is, for sorting and shredding of rags, their product being baled material to be spun elsewhere with new wool into yarn. That year and after, as demand for wool increased with the war, wool manufacturers began adding machinery for the production of shoddy in their mills.[23]

New England mills and factories, many of which had previously been geared, for reasons stated above, toward the production of relatively low-quality goods, became, with the advent of war, vital suppliers of military accoutrements (uniforms, blankets, bags, overcoats). At the same time, as the cotton supply dried up overnight, cotton mills either shut their doors or more often than not transitioned rapidly to wool production. Storerooms for cotton shipped up from the Southern states became depots for woolen rags and clippings. As inventor, entrepreneur, and president of the National Association of Wool Manufacturers Erastus Bigelow described the situation a few years later, "Cotton mills were converted into woolen mills, and new establishments sprang up as if by magic."[24] Profits ballooned, while at the same time widely publicized defects in such items put the industry under particular public scrutiny.[25] Responding to initial reports of inferior equipment arriving at the front, the federal government soon assembled a commission to investigate fraud and corruption in government contracting and procurement.[26] The Committee on Government Contracts, a select committee of the US House of Representatives, was formed on July 8, 1861. Its name henceforth was referred to by the name of its first chair, New York representative Charles H. Van Wyck, representative from New York, one of the states with a large wool and garment manufacturing industry.[27]

There was an immediate sense of relief in some quarters after the establishment of the commission and an exaggerated opinion, perhaps, of its early success, at least judging from an article published less than two weeks later, shortly after the terrible Union defeat at the Battle of Bull Run on July 21, 1861. Its author offered these hopeful and yet chastening words:

> Half the disaster of Sunday is due to the fact that Colonels, Captains and Lieutenants were utterly unfit for their places,—that in many cases they evinced the most shameful cowardice, and in others the most entire ignorance of their duties. It is indispensable that the Army should have officers capable of doing their duty, and in whom the men confide. . . . Officers must be appointed hereafter for their military qualifications exclusively,—not for their local popularity nor their political services. . . . The spirit of the people is now thoroughly aroused, and, what is equally important, it has been chastened and moderated by the stern lessons of experience. Men will be offered by hundreds of thousands for our Army; let the Government take good care that they are properly used, not wasted or left to incompetent and fatal leaders.

Inferior uniforms, along with blankets and overcoats, made of shoddy came in for particular censure:

> The wretched, shoddy and half-hemmed rags that have been palmed off on the men, as clothing, by swindling contractors, will disgrace American soldiers no longer, as every article is now most rigidly inspected, and the slightest defect condemns it. Prompt and regular payment, and means for that purpose, have been abundantly provided for by Congress; and now that the confusion incident to the hasty gathering of such an immense body of men is nearly rectified, and things are getting in regular working order, it is believed that all difficulties on that point will be obviated.[28]

The popular press soon seized on the issue, as illustrated, for example, in figure 2.4. In the accompanying *Vanity Fair* poem and associated illustration “The Dream of the Army Contractor,” the contractor’s nightmare, or “distressing circumstance,” consists of his being forced to wear his own product, his “bliss” of golden dollars “alloyed” as “his clothes were turned into army cloth” from whose “gap[ing] seams” the coins fall “clinkety-clank.”

The sartorial politics that might otherwise have remained in the realm of yellow journalism soon assumed the form of official testimony and permanent public record. Between the latter half of 1861 and the early months of 1862, the Van Wyck Committee conducted hundreds of

Fig. 2.4 "The Dream of the Army Contractor," from *Vanity Fair*, August 17, 1861, p. 77. Illustration for a poem by Charles Graham Halpine, an excerpt of which is: "A distressing circumstance alloyed the bliss of his dream. His clothes were turned into army cloth, and they gaped at every seam. And the golden dollars fell clinkety-clank from tattered trowsers and coat. And when they fell, they burned round holes through the bottom of the boat." (Courtesy of Harvard Library.)

interviews in twelve cities throughout the region. In the course of those inquiries, the textile industry came under special scrutiny, particularly with regard to defective uniforms and blankets. Hearings lasted from the beginning of July 1861 through April 1862, with deliberations going on until July 1862, after which a final report was issued, along with testimony that filled thousands of pages.[29]

The testimony of one Lieutenant Colonel George H. Crosman, a deputy quartermaster general, in March 1862, is representative of the difficulties jointly caused by technical failure and political incompetence (or worse):

Q. Has the government a large depot here for army clothing?

A. A very large depot—the largest in the United States, of which I have sole charge.

Q. What amount of material or of clothing have you on hand at the present time; what is the character; by whom has it been made, and by whom inspected?

A. I have a very large amount of clothing; enough, I suppose, to equip 200,000 men for three or four months, perhaps longer. . . . Of some articles I have enough for double that number of men for that length of time, as for instance, shirts, drawers, stockings, and shoes, form essential articles, which . . . it will never do to be without; we must always have them on hand. . . . [T]here is in the depot a considerable supply of what is called irregular clothing . . . bought at a time when the army standard cloths could not be had, or when what were in the market were held at enormous prices . . . under the superintendence of the regular officer of the arsenal.

Crosman goes on to reveal how several officers of his, and one in particular, had regularly "condemned," for no apparent reason, a "lot of blankets which were perfectly good" while allowing others that were obviously defective. The deputy quartermaster's suggestion as to what should be done with the latter, the "shoddy" goods, is especially revealing.

Q. From your long experience and your knowledge in reference to articles of this kind, what in your judgment would be the best manner for the government to dispose of them?

A. I think the best mode would be, in the first place, to issue them to prisoners of war, if they are to be clothed by the government, and also to the negroes. That would be the first disposition to be made of them; and then they might be issued to prisoners of the army—our own men in confinement for offences.[30]

Fig. 2.5 Shoddy federal blanket used in the Civil War, carried by Peter Thibodeau, a private in Company B of the First Maine Heavy Artillery. Some of the holes are allegedly from bullets fired by a sniper while Thibodeau was asleep. (Bangor Historical Society.) Photo by Hanna Rose Shell.

That defective shoddy blankets were deemed suitable for "prisoners of war" if "they are to be clothed by the government," and also for "negroes," followed by "our own men in confinement for offences," suggests a hierarchy of degradation for which shoddy blankets especially would come to serve as an apt symbol and around which would ultimately appear a cult of remembrance (fig. 2.5).[31] (Shoddy blankets have become highly valued for Civil War reenactors and collectors alike.)

Textile Skin and the "Sinews of War"

Shoddy *qua* shoddy (in the physical sense of industrially processed, ground-down, and respun remnants of unknown previous wearers) was just the beginning. There is an expressivity of weave and wear inherent in textile technologies. Old clothing and related practices of textile renewal have a peculiar vibrancy within this context. The potency inheres in its ability to function as an amplified expression of more general anxieties surrounding the dynamic relationship between industrial mechanization, on the one hand, and embodied (tactile, sensory, and deeply felt) human experience, on the other.

Shoddy became a symbolic and emblematic entity that was always already also material and sensorial; an interface between multiple domains, and in this way akin to another relatively innovative medium of the day—photography. Well-worn clothing's relationship to its (present or recent) wearer repeated the indexical—and yet always questionable—epistemic relationship between a photograph and its subject. A relationship that is indexical derives from a direct, material connection between a thing (for example, a person) and its sign (for example, a finger or footprint). It is generally presented in contrast to an iconic relationship, which refers to a continuity of resemblance (for example, a portrait painted of the person).[32] As described in 1864, for example, shoddy "bore the name and the semblance of a thing," much like a photographic portrait of an individual. And yet, whatever the epistemic relationship, the material one could only weaken: "no more like the genuine article as the shadow is to the substance."[33]

The specific way in which shoddy became at once dematerialized and personified is evident in the slippage of the term from noun to adjective, and from being applied to a thing to being applied to a person. Even during the early stages of congressional hearings, officially held to investigate improper procedures for bidding and favoritism across the board,[34] *shoddy* functioned in the popular press as a shorthand for war profiteering more generally. The trend we saw in "Dream of an Army Contractor" would continue with illustrations like those that later appeared in the *Dollar Monthly Magazine*, which billed itself as the "cheapest newspaper in the world," with the title "Mr. Shoddy[,] having made much Money

Fig. 2.6 Frontispiece for *The Days of Shoddy: A Novel of the Great Rebellion*, by New Jersey–born poet, novelist, and playwright Henry Morford. Published in 1863 by T. B. Peterson and Bros. of Philadelphia. (Courtesy of Harvard Library.)

through Contracts, is invited to an Evening Party."[35] And it would take on a similarly luminous specter in such popular publications as the widely read novel *The Days of Shoddy*, by Henry Morford (fig. 2.6).[36]

Consider, too, the lyrics to "Song of the Shoddy," published in the magazine *Vanity Fair* on September 21, 1861. The song is told from three perspectives—a lieutenant, a quartermaster, and a tailor—over the opening three verses. "I, Lieutenant-Colonel Graham" describes:

> . . . the coats contractors gave us,
> Were of shoddy-cloth of gray,
> Badly made, and badly fashioned,
> Much too large or small for men;
> Only for a day we wore them,
> And they came to pieces then.
> Bad the buttons—bad the breeches,
> Breeches only fit for mending
> O the ripping! O the darning!
> O the tailoring unending.

After condemnation, by both the quartermaster and the tailor, of the terrible quality of the uniform's fabric, the song's final and by far the longest verse is a lament that presents the case of shoddy as emblematic of a grand travesty of epic proportions.[37]

> . . . Oh my country!
> Could it be you did intend,
> Wretches draped in shameful shoddy
> To the battlefield to send?
> Shoddy ripping, shoddy bursting,
> Shoddy rotting in a day.

What might explain this rapid acquisition of complex meanings, shoddy's transformation from a noun into an adjective, the meaning of which ranged from substantial to immaterial, so soon after the introduction of the term into general discourse? Simultaneously known and unknown, present and absent, shoddy was connected to the body of the soldier—indeed, an extension of that body—and yet for all that, degraded, always already broken down, and rooted in an unseen and mostly unknowable past. The *intimate materiality of the unknowable* is ascendant.

Its joint reality and unreality as both a material and an abstract thing was highlighted a few weeks later in an article in *Scientific American* of October 12, 1861, on the subject of what shoddy "is" or "how it is made":

> "Shoddy"—Since the charges, so extensively circulated against a portion of

> our army clothing contractors, of making the soldiers uniforms of shoddy, the word has passed into general use, and has become a synonym for everything that is false. Logwood brandy, a counterfeit note, an untrue statement, a young man who deceives a girl with false promises—are all designated by the expressive term "shoddy." Though the term is applied to everything unreal, the article has an actual existence, and many persons are engaged in its manufacture.[38]

The article's almost obsessive detailing of the processes of shoddy production points, however unintentionally, to the inherent instability of the material reality at issue.

> Woolen rags are $5 and $10 per ton for making shoddy cloth. Fine black scraps are worth $100 to $150 per ton. The shoddy manufacturer passes them through a rag machine, which tears the rag to wool and cleans it of dust. When reduced to soft wool, the shoddy is saturated with oil or milk and mixed with new wool in as large proportion as possible. White shoddy is used in blankets and light colored goods, and the dark description for coarse cloth, carpets &c. The shoddy is the product of soft woolens; but the hard or black cloths, when treated in a similar manner, produce "mungo," which is used extensively in superfine cloths, which have a finish that may deceive a good judge. It is used largely in felted fabrics. Shoddy in the cloth of a coat will soon rub out of the cloth and accumulate between it and the lining.[39]

Another article about shoddy, this one from 1866, attempts to demystify shoddy, to return it from an abstract metaphorical catchall to an actual material good: the product of engineering rather than literary conceit, albeit with a somewhat more positive ending, reflecting, perhaps, the war's perceived outcome. After describing in detail the logistics of manufacture, several of shoddy's high-end uses are highlighted—including fancy wallpaper ("velvet hangings") and thermal insulation felt for houses.[40]

As the war continued, so did the violence of battle, along with the bodily horror of its aftermath. In the absence of clear knowledge about infection's cause or prevention, the ravages of disease—the ill, the dead, and the dying—littered battlefields and field hospitals. The most visible artifact of the carnage were the bodily coverings, blankets as well as clothing, draping the soldiers and often serving as their only shrouds.

Throughout late October, November, and early December 1861, as Quartermaster General Montgomery Cunningham Meigs struggled to equip the proliferating number of regiments, and hence bodies, to be clothed by day and warmed at night, there had been a spate of articles in local and regional daily papers on the "blanket question."[41] Meigs was confronting real and severe shortages, a point he made in his correspondence with the brigadier general in late summer: "The nation is in extremity. Troops, thousands, wait for clothes to take to the field. Regiments have been ordered here [Washington] without clothes. Men go on guard in drawers for want of pantaloons. . . . We must bear the clamor of fools while the country hangs in the balance."[42] "Clamor of fools" or not, concerns about quantity as well as quality were real, because even the best-made uniforms would wear out after a few months with steady use in the field. Blankets posed an especially great problem for Quartermaster General Meigs, who insisted after 1862 that soldiers bring them along at time of enlistment.[43]

Judging from one representative article that appeared in *Vanity Fair* on November 16, 1861, which associates shoddy with a dirt "retentive" of "cheap luxury" along with the "sinews of war," the "blanket question" had as much to do with how to understand and *interpret* such shoddy blankets figuratively as with what to do about their literal shortcomings. *Sinews* refer literally to the material that holds the body together, binding muscle to bone or bone to bone—connective tissues all of them.

Photographs from field hospitals documented a related assimilation of shoddy blankets with images of wounds and infection. In the photograph of a Union field hospital, depicted in figure 2.7, sick bodies become sick clothes, textile "sinews of war," to be sure. In contrast to the scenes of pain in later Civil War films such as *Glory*, we view here a sea of clothing and coverings: gauze bandages and cotton rags, yes, but also, even more so, blankets and torn-up (shoddy) bits thereof, overcoats and felted woolen hats. What is most visible is in the clothes: a vision of dying bodies in deteriorating uniforms, themselves secretly rife with infection transmitting from one body to another. And although the facts about transmission of disease were still generally unknown, a few might have had their suspicions, including Holmes himself, whose father, Oliver Wendell Holmes Sr.,

Fig. 2.7 Union field hospital at Savage Station, Virginia, June 30, 1862. Stereographic photograph (this is one of the two frames) made by James F. Gibson. (Courtesy of the Library of Congress.)

had already published a pathbreaking article on the spread of puerperal fever via contaminated bedsheets as part of a more general contagion theory of disease, an essay that would later become famous.

The sinews of war are on full display in this image. Savage Station was a federal supply depot where the Union army had constructed a makeshift field hospital after the June 27 Battle of Gaines's Mill. After the fateful Battle of Savage Station on June 29, the field hospital was stretched beyond capacity. The moment recorded in the stereograph by photographer James F. Gibson precedes the capture of these men by Confederate soldiers later in the day on which it was taken, June 30. Ultimately, Union forces would withdraw from the area (after losing the battle), and what we see depicted here would have been abandoned—the supplies and approximately 2,500 wounded soldiers. The latter would be taken prisoner, and this after a day of 1,500 casualties on the Union side.

We see the cacophony—corporeal, institutional, and spatial—involved in the attempt to manage illness and injury; comrades are linked as blankets pile up in disarray and swatches of clothing and bandages are exchanged between them. This is a crisis in "connective tissue" played out on the bodies and in the bandaging pervading the scene. The original stereographic version was produced as a large glass negative (4 × 10 inches) with a wet collodion-type exposure. A camera would have had two lenses; one on the right side would capture what the left eye would see, and vice versa; their orientation would be switched for the purpose of making the stereograph card. Figure 2.7 is a print made from one of these negatives. We see men from many different units, with a special representation from a New York regiment (the Sixteenth) and also some Vermonters, "licking their wounds." At nightfall, following the departure of the cameras, the wounded and ill would have been mostly housed inside the makeshift structure—a barn, it seems, along with farm sheds and hastily assembled tents.

We also see the multiple effects of rags—rags that mark where the people are, what was burned when not being used, and what would eventually be generally recognized as fatal sites of infection. There are surgeries under way, or at the very least we can see what looks to be an amputation in process in the foreground. But also evident here is the lack of knowledge about germs as sources of infection: witness, for example, the rags of one soldier's shoddy uniform being used to "clean" things off for another procedure. As the *Vanity Fair* article on the blanket question had presciently intuited, the blankets might be "impeached of dirt," but they remained "retentive" of waste.[44]

Textile skins, whether actual shoddy or merely shoddy-like, were often stripped off dead bodies for reuse.[45] A shortage of goods meant that these, along with shoes, were at a great premium, with soldiers' families on the home front scrimping and pinching for basic goods. *Forgotten, misbegotten on New Hope Battlefield* is the title of the image shown in figure 2.8, as part of the *Photographic Views of Sherman's Campaign*. The very idea of abandonment—its visual resonance in any case in the form of memories, rumors, and then photographs made of battlefields raided of leftover supplies after the opposition's soldiers had retreated, with only the stray boot or blanket scrap left haphazardly—would ultimately be linked to salvaged blankets and other supplies.

Fig. 2.8 "Forgotten, misbegotten on New Hope Battlefield, Georgia, 1864. Battle Field of New Hope, Ga., No. 2," from the album *Photographic Views of Sherman's Campaign* with the caption "Photographed by Mathew Brady." (Courtesy of the National Archives.)

Shoddy and the Body Politic

Shoddy also came to inhabit visual, material, and literary culture, functioning as a stand-in for death, corruption, illness, betrayal, and the ills of unregulated industry—but also for patriotism, sacrifice, comfort, healing, and memorialization. As in so many domains, the relation of shoddy to the larger body politic served to materialize and represent vices and virtues both social and intimately personal. By 1863, shoddy's joint associations with both political and social corruption and with the wounds of battle had coalesced, as in the popular image, probably by well-known illustrator Thomas Nast, from *Harper's Weekly*, shown as figure 2.9. The image resituates the pathos-laden, detritus-strewn landscape of the field hospital photograph within the context of a set of real or imagined "sentimental" relations—economic, social, and familial. These are mediated through the woolen rag cloth manufacturer, the contractor of the center bottom frame.

Fig. 2.9 "Service and Shoddy—A Picture of the Times." From *Harper's Weekly*, October 24, 1863, p. 677. (Courtesy of Harvard Library.)

The contractor and the contractor's wife—sometimes known as Mr. and Mrs. Shoddy, or even just Shoddy and Mrs. Shoddy—was a common trope in a range of late and postwar satirical prose and visual polemic. The characters were often ethnic "others," always ostentatious, callous, unseemly, and generally overdressed.[46] The Mrs. Shoddy figure became especially resonant as a figure of derision and mockery. Superficiality characterized these figures, precursors to the inhabitants of the Gilded Age that Mark Twain would describe in his eponymous novel of a few years later.[47]

In the meanwhile, novels such as the above-mentioned *The Days of Shoddy*, whose elegant frontispiece (fig. 2.6) literally branded an outwardly elegant overcoat with a label of the era, focused on a "shoddy aristocracy" of depraved and vulgar wealthy opportunists. Although Morford's characters weren't named Mr. and Mrs. Shoddy, shoddy predominates as an adjective that refers not so much to articles that are poorly made as to a lack of "class," morals, and courage—in other words, to express derision toward the parvenu or the nouveau riche enjoying their ill-gotten luxuries on the backs of soldiers dying at the front.[48] If there had once been hopes that the new shoddy class would face a swift comeuppance, that expectation had evidently faded: the rich and poor in Nast's image remain wholly separate from each other, leaving the poor, honest, and patriotic without visible recourse. This is two nations, to refer back to Benjamin Disraeli's characterization of 1840s England, distinct from that demanded by the secessionists.[49]

George Isaac Reed's extended poem *The Russian Ball, or The Adventures of Miss Clementina Shoddy* (1863) makes a similar point, employing the language of "Shoddydom" and crass female "Shoddyites." It also includes the theme of foreignness (Russians, in this case), as seen in the placement of the flag and other symbols of the nation on the cover (fig. 2.10).[50] It thereby highlights the connection between shoddy and patriotism and, by negative implication, treason. (Note the American flag in the upper-right corner of the cover, with "shoddy" labeled immediately below it.) Reed's verse describes the luxurious life of one Miss Clementina Shoddy, whose lifestyle is paid for by her father (the contractor Père Shoddy). Reed dramatically juxtaposes Miss Shoddy's many foreign flirtations with the concurrent suffering of American soldiers at the front.

> All Shoddydom flutters with great expectation;
> *The* night of *the* ball to *the* guests of *the* nation.
> The bosoms of thousand fair Shoddyites pant
> with triumph: They're going, while thousands
> more can't.

Reed's verse expands the latter suggestion into a virtual tableau of war:

> I saw at the ball Shoddy sensual and dull,
> Exclaiming: "Ha! ha! I have eaten my full."
> But there at his feet (in my true waking dream)
> Was rolling a swollen and dark southern stream,
> And near it a soldier, who lay by its brink,
> All bleeding and dying and praying for drink.
> He'd fought for the flag, in a skirmish he fell,
> Unknown and alone he was dying—Ah, well,

Fig. 2.10 Cover of *The Russian Ball: or, The Adventures of Miss Clementina Shoddy* (1863). The American flag in the upper-right corner has "shoddy" labeled immediately below it. (Illustration by Robert Buehler.)

> While Shoddy was raising the wine to his mouth,
> The soldier of the North breathed his last in the South.

The verse's feminine associations with shoddy are not wholly negative, however. The poem ends, rather, with the "dream" of a "sweet girl" stitching in a freezing garret "and thinking of him who had gone to the wars." It is of note that that Miss Clementina is depicted as being shoddy insofar as she only appears to be high quality (". . . though *in private* a slattern," she "determines in public at least to be splendid"), whereas, in contrast, her father is (named) "Shoddy" because of his shady business practices. These two failings thus merge, as it were, into the "family" of meanings for shoddy as metaphor in this period.[51]

The explicit contrast drawn between Miss Shoddy and the sweet girl stitching for her soldier beloved calls to mind another literary theme of the time, with similarly ironic parallels to shoddy—the domestic lint and bandage-making that was broadly perceived to be self-sacrificing and even heroic. Such feminine patriotic devotion served as a moral counterweight in the public mind to the corruption associated with shoddy fabric—an instantiation of shoddy's protean ability to materialize opposites, represent antithetical qualities, and mean contradictory things. Consider, for example, the following stanzas from a poem written by Mary E. Nealy for the Indiana Sanitary State Fair of 1864,[52] to commemorate women's contribution to the war effort, and specifically linking such "unstinting" work with the nobility of death on the battlefield:

> It is this that mans us in battle's hour,
> That nerves the arm and gives it power;
> That makes the heart's blood fresher flow,
> And gives to the bosom its noblest glow.
> The women we love, the God on high—
> They well know how we bravely do or die.
> God bless our noble women! [who]
> . . . make the bandage, they scrape the lint[.]

The poem directly links the home-front service of women, whose own "silken dresses [have been] laid away," and for whom "calico dresses are good enough," with the production of another textile of mixed origin: the lint used for bandaging and dressing wounds. It thus juxtaposes the ill-gotten luxury fabrics of the shoddy class with a nobler use of womanly artfulness. In this and related ways, fabric rags and tatters—shoddy, if you will, of a different name—played positive and even poignant roles during this period, not only as dressing that provided aid to the sick and wounded in the field, but also in memorials to the dead and grieving. In this regard, quilts and handmade shrouds similarly associated with women's work and fragments from battle-damaged flags were treated with special reverence.

The home production of lint for bandages and dressings was publicly organized under the auspices of the US Sanitary Commission, a civilian organization established almost immediately after the start of the war in early summer of 1861. It set out, under the leadership of Frederick Law Olmsted, to provide medical support and comfort to Union soldiers. Requests from field hospitals for items in grave need were common; early on, hospital gowns were in especially short supply, and volunteers were asked to track down surplus and used gowns from local clinics. The commission, along with subsidiary organizations, conducted drives to collect old wool, sheets, blankets, and material for bandages.[53] By the autumn, the needs of the army had ballooned, and the commission expanded and merged with various women's relief organizations that aided not only in the collection process, but also in general fund-raising, and various knitting and hand manufacture activities. Although some members assisted nurses at the front, most were engaged in knitting woolens and in the fabrication of lint for use as bandaging.[54] The bandaging itself would end up having an intimacy with punctured and often shattered bodies. As described in a popular contemporary textbook on surgery:

> The general rule to be observed in bandaging a limb are, to begin at the extremity, and apply the bandage most tightly there, and more loosely by degrees as it ascends—to make each successive fold overlap about one-third of the preceding—to keep the bandage close to the limb, and unroll very little of it at a time—and to double it on itself on parts (such as the calf of the leg) where it would not lie smoothly otherwise.[55]

As in figure 2.11, the bandage thereby became a second skin delicately applied in the hands of a skilled field surgeon.

Lint (a material made from the fuzz left on used cotton and linen) was a product in especially high demand as the substrate for such precision maneuvers. Women "picked it," either as loose threads or as the fuzzy material scraped off fabric with a knife. Although machine-made (or patented) lint was available, especially in the North, handmade lint was especially prized, in part for its moral associations, as a labor of womanly love. (Such associations are especially poignant in retrospect, given the

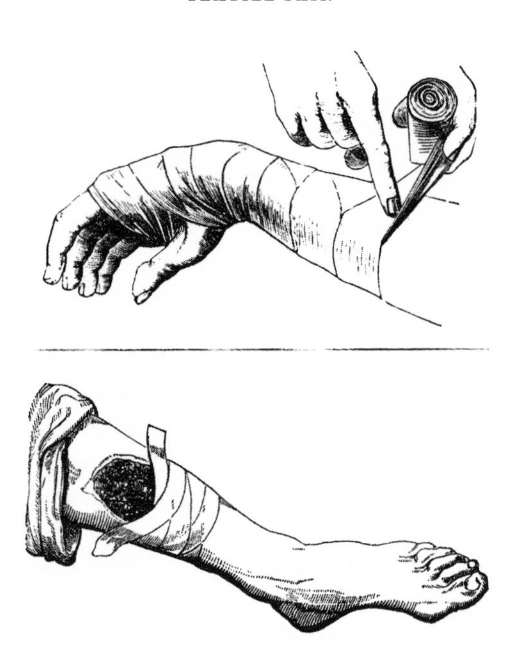

Fig. 2.11 Illustrations from Robert Druitt's *Principles and Practice of Modern Surgery* (published by Blanchard and Lea in Philadelphia). *Top:* Bandage for the forearm (1865 edition, p. 608). *Bottom:* Dressing the wound, indolent ulcer (1867 edition, p. 89). (Collection of the author.)

unsterile nature of lint, along with the bandages, often wet, that usually covered it.) One female volunteer nurse wrote in her memoirs about a "lint and bandage mania" on the home front and even at her field hospital, while at the same time lamenting the poor quality of shoddy uniforms and blankets only increasing the seriousness of illness, with soldiers "wearing only shoddy uniforms and protected only by shoddy blankets and the result was a frightful amount of sickness."[56]

Although unlike shoddy proper, lint did not consist at least partially of woolen fibers (linen and cotton were preferred), medical journals of the period sometimes referred to lint in its "coarse" form as "shoddy,"[57] adding to the metaphorical slippage implicit in poems like Walt Whitman's famous "The Wound-Dresser" (1865), the very title of which draws attention to the connection between clothing, rag-based bandages, perforation of the skin, and death. The bottom part of figure 2.11, with its focus on the relationship between the rending of skin and its textile recompense, articulates this connection in a manner both poetical and pedagogical.[58]

Photography and the "Harvest of Death"

The Civil War also gave rise to the proliferation of another circulating medium of representation, photography, one that gave rags new visual resonance captured in arresting images of suffering and deceased human bodies with their shoddy-derived (and often shoddily manufactured) uniforms. Photography, too, shared with shoddy a bit of the uncanny, having multiple resonances with the source of the representation. A series of celebrated photographs, to whose more detailed analysis I now turn, enabled the displacement of both corporeal horror and sentimentality onto an alternative and potent textile skin through physical connections, both tactile/tacit (indexical) and visual (iconic).

In distinguishing our current way of looking at such images from how they were likely to have been seen by earlier audiences, it may be helpful to recall the image of Oliver Wendell Holmes Jr. (fig. 2.1), for whom, even sixty years later, the Civil War remained a vivid personal memory.[59] His deliberations on *Weaver v. Palmer*, with their explicit identification of shoddy as an agent of infection, were still more than half a century away when he and the rest of the Massachusetts Twentieth Volunteer Infantry arrived in Gettysburg, Pennsylvania, site of the famous brutal battle, conducted between July 1 and July 3, 1863, in which thousands perished, including many in his own unit. How are such scenes likely to have appeared to battlefield soldiers such as Holmes, and how remembered? And how (if at all) might what they saw have differed from what confronted the professional traveling photographers who recorded these scenes with the aid of a variety of reproductive devices? Finally, how, in turn, might each have differed from what was "seen" by the readers of the later "sketch books" and other commemorative materials in which final versions of such photographs were published at the end of the war?

Just one day into the fighting at Gettysburg, the landscape was already littered with corpses—bodies of the dead and dying—documented en

Fig. 2.12 "Gettysburg, Pa. Confederate dead gathered for burial at the edge of the Rose woods, July 5, 1863." Photograph by Alexander Gardner. (Courtesy of the Library of Congress.)

masse in photographs by Alexander Gardner that would become iconic images of mass, partly mechanized warfare. In figure 2.12, we see the dead on a battlefield, and also—in the upper left-hand corner, part of the carriage being used as a portable darkroom by the team of photographers.

Gardner, along with Mathew Brady, who commissioned much of Gardner's work, is perhaps the most widely known among a circle of famous Civil War photographers, including Timothy O'Sullivan, James F. Gibson, and George N. Barnard, who together provided the best-known images of large-scale military engagements, following upon the development of new photographic technologies that made such images possible. Their photographic work (in everything from daguerreotype to large format) of the

Fig. 2.13 "Federal dead on the field of battle of first day, Gettysburg, Pennsylvania." Print made directly from negative in the form of a color film transparency. Photograph by Timothy H. O'Sullivan. (Courtesy of the Library of Congress.)

various campaigns not only created the most widely recognized images of the war, both at the time and for posterity, but also transformed the way wars were viewed and remembered more generally. The American Civil War was the first major war to follow the inventions of a set of interconnected image-making technologies: the stereo camera, the ambrotype, and the wet collodion process. Gardner, who was at the time working with O'Sullivan and Gibson, was the only traveling photographer to reach Gettysburg before the bodies were buried or embalmed or otherwise prepared for shipping elsewhere (fig. 2.13).

From the perspective of those on the home front, the work of these photographers provided an intriguing window into (and a certain way of framing) the events of war. Since photographs could not yet be reproduced easily in publications, engravings based on the photographic plates appeared in widely circulated publications such as *Harper's Weekly* and

Vanity Fair. During the war, photographic plates and stereographic views were exhibited in private galleries and offered for sale through high-end catalogs. For example, "The Dead of Antietam" exhibition at Mathew Brady's New York gallery displayed both Gardner's and Gibson's stereo photographs. Meanwhile, Gardner's gallery released the "Catalogue of Photographic Incidents of the War" series in September 1863.

Things were different by the war's end. After 1865, many of the single-frame photographs, most of them never before viewed by the public or even printed, were incorporated into elegant "historical" volumes: most famously *Gardner's Photographic Sketch Book of the War*. Through such sketchbooks, narratives constructed in prose around particular images came to be accepted as historical fact, the photographs gaining the status of documents of truth par excellence.[60] The repetition of the content of the images through multiple, slightly different camera angles ironically served to support and reinforce the idea of a kind of ultimate "photographic truth," acquired from an accumulation of slightly different perspectives. Yet these scenes *were* staged. The photographs of Gettysburg made by the most notable of the Civil War photographers—including Gibson, Gardner, Brady, and perhaps others—have been discussed at length by a range of historians of photography, art, and the Civil War.[61] Much has been made of the fact that many of the bodies we see in these images were in fact posed by the photographer(s) and assistants in what have been called "doctored presentations" (as if such a thing as an "unmanipulated" photographic image, whatever that might mean, were actually possible).[62] There can be little doubt that bodies were often moved and redressed in different uniforms, that Confederate soldiers were sometimes purposely represented as Union dead (or the reverse), and that shooting locations were at times misidentified, either accidentally or intentionally. Some have argued that such misrepresentations were mainly the result of the limited number of exposures possible, given the available technology, encouraging photographers such as Gardner to seek to connect a "plausible image" with a "convincingly written narrative."[63]

I would contest the claim, however, that the mere application of narrative on top of image is what gives "reality" to the depiction. And references to "rhetorically convincing effect" are similarly vague and incomplete, as are the appeals to "ideology of Civil War photography" on the part of some contemporary scholars.[64] More broadly considered, such photographic depictions, which were often posed with a view to certain allegorical and poetic associations, render palpable a peculiar merging of the corporeal body and the textile uniform. Dead bodies appear as strewn bits of rags, cloth skins of death after the ravages of Gettysburg. Bodies appear as textile heaps in the foreground; a (no doubt) shoddy federal blanket serves as an impromptu shroud. Likely as not, it was embroidered with a US seal by one of the many manufacturers (contractors) of low-quality goods that would (or were said to) "fall to pieces" on their own, even without the strains of battle.

Thus, the image (also) registers through the textile medium itself—the particular blend of iconic and indexical qualities it offered and offers. The apparent finality of the published images in *Gardner's Photographic Sketch Book* belies more than the multistaged manipulation and fluidity—chemical and metaphorical—involved in the original processes.[65] Fabric posing as skin belies, one might say, the distinction between artifice and mimesis, fiction and documentary—it is inherently simultaneously dead and undead, human and technological. "Used," as it were, such textile skins become indelibly marked with the human corpse, the body.[66] And this is true regardless of the intentionality or awareness of the photographer, the writer, the marketer, even the gallerist, the stereoscope maker, or the contemporary viewer.

Note the way in which textiles are both like and unlike other technologies pictured in these photographs that were also widely employed during the Civil War—one of the most notable features was the use of interchangeable parts. Rifles, such as those shown in figures 2.14 and 2.15, are the prime example here. Such parts were manufactured so as to be identical for all practical purposes both before and after initial use. The Civil War was the first on American soil where such manufacture was possible and the manufacture of interchangeable parts was carried out on a massive scale. Whereas clothing and uniforms, even blankets, may well begin as interchangeable (in specific sizes, for example), they are much more immediately affected and reshaped by wear (that is, through use) than the rest of the array of such technologies—guns, in particular.

In photographs such as *Sharpshooter's Last Sleep*, one is struck by how fresh and clean, practically brand-new, the sniping rifle is; one could

Fig. 2.14 *Sharpshooter's Last Sleep*, from *Gardner's Photographic Sketch Book of the War* (1865–66). (Courtesy of the Library of Congress.)

almost walk off with it. By contrast, the dead soldier's clothing is more decomposed than the man himself. The rags in the foreground are no better, and the hat blown off his head is ratty, indeed (fig. 2.14).

Several of the photographs in Gardner's album employ this same corpse and rags, along with the same US-issue rifle, which is itself probably a prop. In *Home of a Rebel Sharpshooter* (fig. 2.15), as well as *Sharpshooter's Last Sleep*, our eye is drawn to the position of the gun, which is the same in each (hence probably the result of deliberate placement); while the body, judging by its clothing, is actually most likely that of a single infantryman.

In *Home of a Rebel Sharpshooter*, a photograph of what we now know to be a scene posed by Gardner and his assistants, Gardner's text places special emphasis on how the old clothes are arrayed, as if pointing, albeit inadvertently, to its deliberate nature. Gardner has taken the body from

Fig. 2.15 *Home of a Rebel Sharpshooter*, from *Gardner's Photographic Sketch Book of the War* (1865–66). (Courtesy of the Library of Congress.)

the initial scene, in which the clothing practically assimilated visually into the earth, as the earth, just as would have occurred if the sharpshooter was doing a top-notch sharpshooter job. He took these materials—blanket in the foreground, body, uniform—and dragged them all a long way (between seventy and eighty yards, it is thought),[67] and then resituated them, paying particular attention to the placement and exact arrangement of textile remnants of this soldier's living days.

It seems likely that the body was chosen because of how clean it remained, even a few days after the battle; visually, in its whiteness and apparent pristineness, it meant that all the pain, suffering, et cetera, could be projected, so to speak, into the decomposition of the surrounding textiles. In the absence of the true ghoulishness of an actual bloating corpse, rotting head, and so on, the body itself could be conceived as simultaneously "lifelike" as a marble statue and as if embalmed in a manner that calls to mind the domestic memento mori photographs that would follow in the coming decades. Things have been laid out just so, in precisely the way that would ultimately allow for the construction of the rhetorical narrative describing the scene: that he had "laid down upon his blanket to await death."

Clothing here plays a vital role as simultaneously iconic and indexical (both looking like and connected by touch) in relation to both the body itself and its passage from life to death (a trace of it, embedded into the weave, the wear, the blood, and so on). Photography occupies a position between icon and index; it both bears a visual (iconic) resemblance to the referent and carries the footprint (indexical trace) of it. (The photograph's duality in this regard might be compared to a threadbare T-shirt whose contours conform to the very body from whom it never wanders, and also whose wear and tear have resulted directly from that body's physical impact on it.) Processes of fixation and layering of emulsion to fibrous paper have been critical to photography from its earliest incarnations two decades before the outbreak of the Civil War. Purity and timelessness are features of its material genealogy.

Gardner adds, in retrospect, that there was no means of judging how long the soldier had lived after receiving his wound, "but the disordered clothing shows that his sufferings must have been intense." Gardner points to the uniform and blanket again as signs of both the state of the dying soldier's feelings and the passage of human time. He describes a return visit to the site four months after making this exposure, to find "the moldering uniform" amidst a scene in which most everything else was static; where the bones were bleached, skull undisturbed, the blankets and textiles continued to be the link between the deathliness of forever and liveliness, fabric once connected to a fleeting, once unrecognized, and now absent human form.

In this narrative, one is left looking at the present photograph in such a way as to focus on the uniform and stony pillow (a coarse blanket wadded against an enormous boulder) as being key spots that ground us in the image. The moldering is the link to the body of death, to the identity of the lost life, and it rots away to become the basis for another kind of living thing. The rifle, meanwhile, propped in the scene here, is not a sharpshooter's rifle, but the photographer's own. Indeed, it is not the rifle, but the textile materiality of the decaying uniform and blanket as pillow

Fig. 2.16 *A Harvest of Death, Gettysburg, Pennsylvania*, from *Gardner's Photographic Sketch Book of the War* (1865–66), plate 36. (Courtesy of the Library of Congress.)

that root this image, that tie it to fact, to life, to the scene at hand; these are the shredded textile skins that are also the sinews of war, its fabric.

In the Gardner photographs, the rags bear the brunt of the injury and the atrocity, leaving the pure whiteness and apparent innocence of skin to last. It is the rags that are punctured, having been perverted by war, just as they had been already in the very act of their production. Henceforth, following Siegfried Kracauer's formulation, the photographs "wander ghost-like through the present."

The uniform and blankets depicted throughout are the left-behind, the remnant—the body *politic*, as opposed to the body, left in tatters for all to see. The fabrication of the uniform, and its very taking of shape, stands in for the corruption and moral impurity of the war on the Northern side. The nobility of the corpse becomes paramount, the body appearing virginal, as if untouched—its purity only reinforced by the shoddy other. The uniforms are opposite, whether those inscribed in Gardner's photographs or hiding in the chests of local historical societies, wool perverted and adulterated by the blending of materials, races, and histories.

Some stereographs released for sale in 1863 were alternate exposures of scenes and imagery that later appeared in the influential postwar memorial that took the form of *Gardner's Photographic Sketch Book*. Several other images made in Gettysburg are worth examining as part of our inquiry. In particular, we can look at the way the scenes were captured and distributed initially (via stereograph, as part of an "Incidents of War" series for sale by mail) and then via the narrated postwar format of the book. One series of particular note is associated with an image that Gardner called *A Harvest of Death* (fig. 2.16), composed of variants on the negative for what had originally been titled much less metaphorically (see fig. 2.13).

As was also the case in *Sharpshooter's Last Sleep*, a single human corpse is here used to articulate different scenes and narratives. Note that the stereographic images (I am showing only one side of each) support a variety of complex readings. Consider how in figure 2.13 there is much fading, dust, imperfection in the emulsion and illegibility in what appears to be a blown-out background and horizon. *Sharpshooter's Last Sleep* has a similar aura of ambiguity, not so much on account of the quality of the image (in terms of exposure or development), but rather as an effect of the scene itself: a blurriness of foreground and background, textile and foliage, dead and alive, as if to suggest that there is a common pattern to things, as if all is settling out into a single plane. The effect is auratic.

The version in figure 2.13 reveals something of the way in which the image is built, letting us see the traces of masking and get a feel for the various chemical processes at work—a kind of surface quality, showing us the layered materiality of both the exposure and printing processes. In the version from *Gardner's Photographic Sketch Book of the War* (fig. 2.16), by way of contrast, these processes are no longer visible. As Gardner himself puts it in the preface to the *Sketch Book*: "Verbal representations of such places, or scenes, may or may not have the merit of accuracy; but photographic presentments of them will be accepted by posterity with an undoubting faith."[68] Gardner continues in the text that accompanies the image:

> Through the shadowy vapors, it was, indeed, a "harvest of death" that was presented; hundreds and thousands of torn Union and rebel soldiers—although many of the former were already interred—strewed the now quiet fighting ground, soaked by the rain, which for two days had drenched the country with its fitful showers. A battle has been often the subject of elaborate description; but it can be described in one simple word, *devilish!* and the distorted dead recall the ancient legends of men torn in pieces by the savage wantonness of fiends. Swept down without preparation, the shattered bodies fall in all conceivable positions. The rebels represented in the photograph are without shoes. These were always removed from the feet of the dead on account of the pressing need of the survivors. The pockets turned inside out also show that appropriation did not cease with the coverings of the feet. Around is scattered the litter of the battle-field, accoutrements, ammunition, rags, cups and canteens, crackers, haversacks, &c., and letters that may tell the name of the owner, although the majority will surely be buried unknown by strangers, and in a strange land.

Thus, through the technological lens of the camera and the presence and dissemination of photographs comes a perceived multiplication of the availability of authenticity, supplementing and supplanting memories and discourses in a way that elides the artificial and the real, invention and documentation.[69]

Like the photographs in *Gardner's Photographic Sketch Book of the War*, after the conflict was over, textile skins functioned in much the same way.[70] They appeared as a surplus at the center of authenticity, binding the body and technology, presence and absence, iconicity and indexicality in seemingly determinate ways. Such was the case of the shoddy blanket. As we have seen, for the living soldier, it had offered both the promise of real, physical support and evidence that the corruption in the production of its material fabric belied that promise; and on the home front, it had supported a multiplicity of different and contradictory claims about its material status and abstract significance. After the war, however, the shoddy blanket—while bearing the traces of life intermingled with death, technology intermingled with the earth, and despite its related reputation as a material particularly associated with death, rot, and impermanence—came to be treated as an equivalent of the "embalmed" permanency of a photograph.[71] One might consider it, to use the language of film theorist André Bazin, "a transfer of reality stemming from the 'mummy-complex.'"

Curated as relics of the war and later collected by antiquarian enthusiasts and reenactors, shoddy blankets, like the photograph in figure 2.13, elide their equivocal nature in an appeal to a narrative that they are said to authenticate. Such a blanket exists in a museum collection at the Bangor Historical Society in Maine, made of union cloth (shoddy wool and cotton blend) and belonging to a private in Company B of the First Maine Heavy Artillery (figs. 2.5 and 2.17). The extreme fragility of the original blankets (on account of their age and the inherent shoddiness that is their legacy) makes them in relatively short supply, compared to many other Civil War collectibles, including textiles, and they accordingly command a steep price, as collectors have assured me.[72] When they do come onto the collectibles market, they are often sold with "authentication" in the form of a letter, either written by the soldier or a descendant.

The attached information is evocative of the narrative that is likely to have accompanied its transmission to later generations:

> Blanket is light olive green with dark olive green stripes up the sides. In the middle of the blanket U.S. is embroidered. The weave is twill. This was carried by Peter Thibodeau of Orono, Maine. The grandfather of Marion Atwell Alton and was carried as part of his blanket roll throughout the Civil War. Holes in the blanket are from bullets fired by a sniper when he was asleep one night at which time he received the only wound of the entire service period . . . [which was] in the form of the loss of one little toe.[73]

As it happens, however, this description belies Thibodeau's actual experience in the war. His unit (he had enlisted in the Eighteenth Maine Volunteer Infantry Regiment, whose name was changed in 1863 to the first Maine Heavy Artillery) suffered the greatest number of casualties in a single day of any unit throughout the war and was also the unit with the most

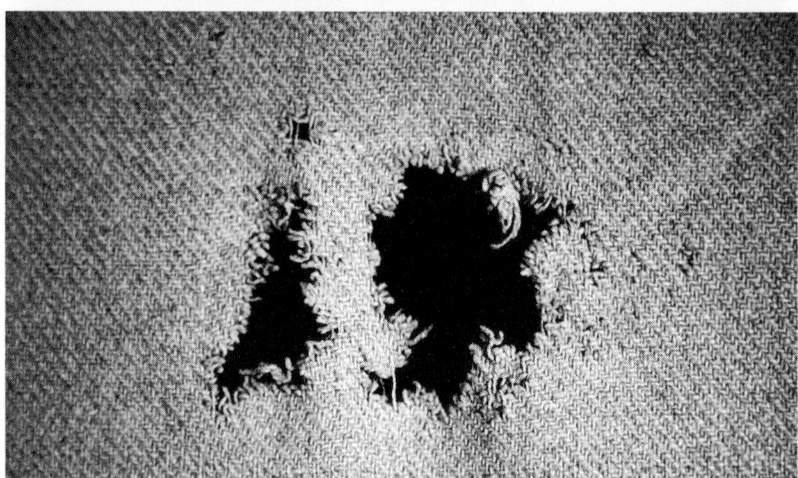

Fig. 2.17 *Top*: Blanket as corpse. *Bottom*: Bullet-hole perforations. Peter Thibodeau's shoddy blanket, at the Bangor Historical Society. Image by Hanna Rose Shell.

soldiers killed overall, with over two-thirds of them killed on a single day during a failed charge against the Confederates in the Siege of Petersburg.

On Shrouds and Shoddy

An unidentified and unmarked ambrotype of two Union soldiers wrapped in their blankets and holding a single flag elides a similar ambivalence in the pathos of their representative anonymity as soldiers of the North in the bloodbath that was the Civil War. Do they sit there covered with their blankets owing to the cold, or for sheer psychological comfort, or as a trying out ahead of time of their own shrouds (fig. 2.18)?

Shoddy blankets, and photographs made of men in their rapture, lingered on after the war. As Susan Sontag remarked, "All photographs are memento mori."[74] These two soldiers bear the traces of life intermingled with death, calling to mind Bazin's provocative "note in passing" in his "Ontology of the Photographic Image," in which he observes how that holiest of shrouds (that of Turin) "combines the features alike of relic and photograph."[75] Bazin compares the shroud directly to photographic images. In the photographs discussed in this act—this ambrotype or *A Harvest of Death*, for example—there is a double functioning of the images, for they provide both the shroud as photographic skin and the shroud as "iconic" representation of the form of the clothes of the dead.

The appropriate comparison, then, is not to the Shroud of Turin itself, but to the controversial photographs made thereof. I refer to the long-disputed evidentiary status of the photographic trace, of the allegedly authentic shroud, as in the widely circulated photograph of the Shroud of Turin by Secondo Pia (fig. 2.19). Though the "degree of separation" is one layer further removed in the case of the photographs of any of the shrouds we here regard, the sense of closeness to the revered original (whether fallen soldier or holy figure) is nonetheless transmitted. As the shape of the body imprints itself onto the fabric, the fabric transmits this corporeal imprint directly to the photographic emulsion.

One might, by way of comparison, and prompted by the Bazin's potent formulation, consider precisely the way in which this shroud was rendered as photograph in the decades following the Civil War, a "textile skin" par

Fig. 2.18 Mysterious ambrotype, made between 1863 and 1865, donated by Waterbury Companies to the National Museum of American History. It came in with a large accession in March 1975 and has neither maker's mark nor any identification of the two soldiers. (Courtesy of the National Museum of American History.)

excellence whose status as such came to be widely shared only upon the distribution of its first photographic tracing, in 1898. In one sense, it is distinct from a shoddy blanket, quite the opposite of a shoddy blanket—linen, pristine in its first application, as opposed to formed from countless unknown others' clothing. And yet, after the photograph's fabrication and then the image's widespread circulation, its origins also became evocatively subject to debate, this aspect of it in itself lending the shroud's link to a fallen man, a presence especially unknown. After all, the photograph

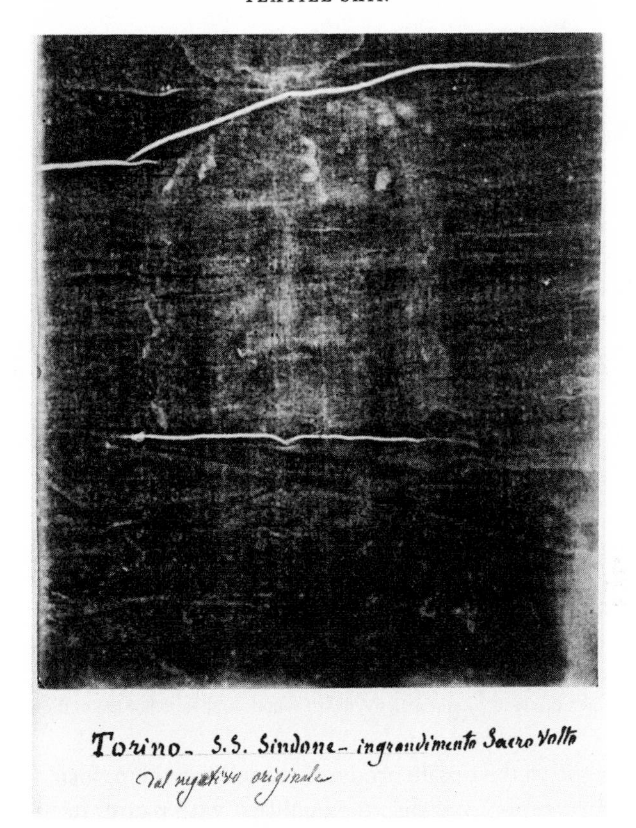

Fig. 2.19 Secondo Pia's 1898 photograph of the Shroud of Turin. (Musée de l'Elysée, Lausanne.)

bore witness to the actual material, as Gardner's images did for the shoddy shrouds on the battlefield, and yet in so doing, the photograph made more omnipresent the question of whose body the cloth had actually brushed against in life and death.

As Oliver Wendell Holmes Sr. had already observed in 1859, with the advent of photography (whose images he likens to "skins being shed" like "cortexes" or "the bark of trees"): "Form is henceforth divorced from matter." Matter "as a visible object is of no great use any longer, except as the mold on which form is shaped. Give us a few negatives of a thing worth seeing, taken from different points of view, and that is all we want of it."[76]

Fig. 2.20 "Our released prisoners at Charleston, S.C., exchanging their rags for new clothing." Illustration from *Harper's Weekly*, January 14, 1865, p. 29, based on a drawing by war artist William Waud. (Collection of the author.)

In rags and in the textile products made from them, form and matter likewise part company as discrete qualities, with matter itself becoming formless and form materialized in multiple ways. Members of his son Oliver Wendell Holmes Jr.'s unit, the Twentieth Massachusetts Volunteer Infantry, had been captured and imprisoned for many months when they were released from Libby Prison by their Confederate captors. A series of etchings appeared in *Harper's Weekly* on January 14, 1865. In a key image from this series (reproduced here along with the original study sketch in figures 2.20 and 2.21), the release of a large group of Union prisoners of war is depicted, including some of Holmes's (now much-depleted) unit.

The etching, based on a sketch by British artist William Waud, commemorated the triumph, depicting the released men mid-dress—as their bodies lose their "old rags" and are given a new set. The enhancements to the sketch made when it was transferred to the etching plates play this up—the raggedness and the "textility" of things becoming especially cen-

Fig. 2.21 Drawing in pencil and china white-and-black ink wash by William Waud, December 1864, and the original sketch for the *Harper's Weekly* plate (fig. 2.20). Inscribed on the back of the sketch: "The figures on the right are coming in with the new clothing, in the centre pitching the old rags overboard, & going out on the left to get their rations." (Courtesy of the Library of Congress.)

tral. There is a deep symbolism here that is also a palpable materiality.

These images imbue clothing with major symbolic importance—that what the prisoners had been wearing, rags themselves made of rags, were somehow a part of their (disgrace and) subjugation.[77] In a deep sense, it seems that even more than nutrition, it was clothes—real clothes—they sought, which connoted personal liberty, the shackles being off, so to speak. The men disrobe, and some of the rags and old blankets end up piled in a heap; others are thrown overboard. All this happens, indeed, as the US flag flaps unflappably (virgin wool as it is). Renunciation of the rags, the shoddy, can be seen as an attempt to throw off the shackles of the Confederacy and simultaneously the deceitfulness of government contracts; it is as if the two become entwined in the very rags themselves. We witness the "throwing away" of form only, exchanged for a supposed "real thing"—from rags to clothes.[78]

A comparison with the original Waud drawing on which the etching was based suggests the magazine's interest in playing up the artist's emphasis on the piles of rags as well as the flag blowing in the wind. Juxtaposing the original drawing with the etching serves to emphasize precisely these features as at the crux of a heroic narrative in journalistic form.

The flag depicted in the image poses a different kind of textile issue, albeit one similarly deeply connected to issues of both individual and national identity. If the blankets could be assumed (almost without fail) to be made in the United States and thought likely to be of low quality and/or with some shoddy components, the inevitably woolen flag could be virtually guaranteed to be neither composed of shoddy nor American made. American flags up until almost the end of the Civil War were made not from American but instead from British worsted wool fabric (specifically, the lightweight wool worsted known as bunting).[79] The first major speech before the newly formed National Association of Wool Manufacturers, in September 1865, complained of an "absolute discrimination in favor of the English worsted manufacture," one result of which was that "we actually make no bunting." "To our shame be it spoken, all our flags are grown, spun, woven, and dyed in England; and on the last 4th of July, the proud American ensigns which floated over every national ship, post, and fort, and every patriotic home, flaunted forth upon the breeze the industrial dependence of America upon England."[80] Among the audience was likely to have been General Butler, whose recently incorporated United States Bunting Company would soon produce the first American made flag to grace the US Capitol. The ambrotype shown in figure 2.18 similarly juxtaposes the same two problematic textiles—one whose low quality made it a public embarrassment, the other linked symbolically and materially to the very nation from which the United States had once battled to free itself.

ACT III

Lively Things

Q. Now there was some talk about some of the [rag] bales being offensive, and they were opened and I think you said there was some suspicion of their being a germ of Japanese life there in the shape of a baby; what was the matter with those rags?

A. I don't know.

Q. You pulled the rags apart, and as they dried everything about them disappeared?

A. I am not positive about that; I know we examined them very carefully and found no baby or rats; we thought it was rats. it smelt like decomposed flesh.[1]

FROM A RAG DISINFECTION CONSPIRACY CASE HEARING AT THE SUPREME COURT OF NEW YORK (LOCKWOOD AND MCCLINTOCK V. BARTLETT ET AL.)

At the 1885 International Sanitary Conference in Rome, Robert Koch and Rudolf Virchow were talking about rags. Koch, widely considered the founder of modern microbiology, and Virchow, the "father" of modern pathology and social medicine, debated how to deal with cholera, and whether proper regulation of the fibers used for shoddy might be an effective method of staving off the pandemic. Koch argued that while the disinfection of rags *was* physically possible, doing so was purposeless without the regulation of "human intercourse" itself. For disinfection of a specific bale to be a success, Koch stated, "the rags may be thoroughly dried before they are packed." However, at the level of public health, the regulation of rags was inseparable from that of living human bodies; as he continued, "but it seems to me to be of little use to prevent the importation of rags if human intercourse is to be allowed." Virchow responded by complicating the very essence, the epistemology, the thingness, of rags. He responded to Koch: "The term rags is not a technical one, like 'dirt.'"[2]

*

What kind of thing is a rag? How does it come to function as both the substrate of shoddy and the conduit of such a range of anxieties about disease and the status of commodities as it moves between bodies, forms, and nations? A rag is a talisman and, assembled en masse, then ground to bits, an abject and primal matter. The tensions embedded in shoddy lent it an expressive capacity, granting it a peculiar and particular potency, even apart from the issue of immigrants with which it was so often bound up.

Rags and the shoddy produced from them existed both liminally—between people's own and others' skins, on the one hand—and substantially—as articles of commerce and commodities, on the other. On account of this duality, they evoked fears and instantiated puzzles of distinguishing between person and thing, between native and alien, self and other. The puzzle of rags emerged early on; they certainly had been around a good millennium before shoddy's mid-nineteenth-century emergence. From the Middle Ages onward, rags in many ways embodied the mysteries of transition and transmission: where they came from, where were they going, to whom they belonged, what they would become. The shoddy material made by grinding them up complicated and multiplied these questions of betweenness—between waste and commodity, between person and thing, even between real and substitute, authentic and artifice.

Since well before the nineteenth century, rags had been associated with concerns about disease, and in the first half of the nineteenth century, in aspects both material and discursive, they were caught up in debates about competing theories of disease—one associated with the Galenic theory of airs and humors, the other based on a theory of contagion via the transmission of discrete entities, whether inert or living things. More and more explicitly, old clothes and the rags they became, and from which sometimes new clothes were made, became a source of interest to people in the growing field of public health and in the laboratory trying to evaluate theories of disease and cures for it.[3] Rags occupied a singular space that could support both theories of how diseases are transmitted. They might be employed as impermeable shields against unhealthy airs or recognized as permeable collectors of them, and they also might be identified as the actual physical agents by which disease was transmitted.

Rags puzzled, confused, or otherwise disrupted what might seem to be a clear distinction between contagion and humor-based theories of disease. Even as germ theories of disease came to be increasingly accepted by the late nineteenth century, they posed problems both theoretical and practical for those seeking to prevent epidemics, disinfect the agents that served as their vectors, and safeguard the public from their depredations. Rags were intimately connected to people. As such, the international flows of rags and immigrants commingled in the popular imagination and public discourse; rags became "lively" alongside increasing fears of ethnic contamination, of "human intercourse."

Seen in these terms, even as the role of rags as disease agents seemed to be clarified, the long-term association with germs, bandages, and sick bays lent shoddy and its rangiest constituents a new and widely exploited potential as cultural resource. Shoddy—rags collectivized, anonymized, and hidden in articles of daily life as nothing else could—supplied a variety of domains, from the visual and textual rhetoric of commercial interests in industrial management and governmental regulation to the early twentieth-century avant-garde. Existing in between so many domains, the protean quality of shoddy and the rags that composed it made it available in ways that could serve multiple and contradictory social, economic, and cultural ends.

Miasma and Contagion

Humoral theory and the concept of looming, lingering, and circulating vapors carrying disease, understood as "miasmas," held the greatest sway in medical understanding of how diseases spread well into the nineteenth century. This theory of disease had emerged in ancient Greece in the work of Hippocrates and Galen, with "miasma" referring to a dangerous pollution or contagious emanation. The universe, it was thought, consisted of the "airs" of fire, air, earth, and water; the human body, meanwhile, contained corresponding "vital humors": blood, phlegm, yellow bile, and black bile. General human decrepitude, as well as particular forms of illness and ill health, were understood as the direct result of imbalances among the humors, in association with specific "airs." In the context of nineteenth-century medical knowledge, as inherited from the preceding centuries, miasmas were floating environments—often characterized by a lingering, cloudy nature, and one that might shift slowly from one region, or one city, or one group of people, to another—and were connected to toxic shifts in internal humoral balance.[4]

By the late nineteenth century, with the vindication of the biomedical work of Virchow and Koch among others, the competing idea would be the celebrated germ theory, which suggested that disease was focused on discrete entities. And that these were either inert or living things, whether or not any germ in particular, that might be effectively located. Germ theory also had a history, with germs previously having been referred to as "seeds of contagion." Here the seeds of contagion, while they existed in the bodies of the ill, did not travel on their own. Rather they attached to so-called *fomites*, from the Italian word for tinder or "touch-wood."[5] In 1546 an Italian physician and poet, Girolamo Fracastoro, had proposed a theory of disease based on this fomite idea, to counter the Galenic model, proposing that spores, tiny living things that were themselves corrupt, physically transferred disease from one individual to another. But he theorized that the "seeds" were hard-pressed to do their work alone; "mobile fomites" were the most likely culprits. "I call fomites such things as clothing, linens, etc., which although not themselves corrupt, can nevertheless foster the essential seeds of contagion, and thus cause infection."[6]

The great plethora of textile items—finished goods old and new, as well as their material, animal, and vegetable constituents—became part of medical efforts and social fears. Clothing could be considered as a potential vector for, or alternately as armor warding off, disease. Efforts to control plagues since the late Middle Ages had included quarantine and separation of the diseased, either through their banishment or by shutting down an afflicted town to prevent exits or entrances.[7] As plague continued to devastate communities throughout Europe, clothing was seen to serve a similar function. It was assumed to provide a possible barrier to disease transmission. Clothing, unlike a fortified wall, still theoretically allowed movement into and out of infected areas. As described in a poem inspired by the Great Plague of London:

Plate 1 "Virgin Wool, Shoddy Temple" (detail from stained glass, Batley Central Methodist Church). Photo by Hanna Rose Shell.

Plate 2 "Masses upon Masses" (shredded rags, crimson-stained wool, and plant growth). Photo by Hanna Rose Shell.

Plate 3 "A Shoddy Typology." Image by Hanna Rose Shell.

Plate 4 "The Devil's Teeth." Photo by Hanna Rose Shell.

In places by the plague appalled
Their hats and cloaks, of fashion new
Are made of oilcloth, dark of hue
Their caps with glasses are designed
Their bills with antidotes all lined
That foulsome air may do no harm,
Nor cause the doctor man alarm.[8]

Medical treatises featured a range of apparently impermeable and visually improbable costumes as frontispieces. Rather than a woven material as the appropriate protective layer, a more literal second skin seems often to have been applied: interlocking layers and fittings of Morocco leather, along with oversize projectile leather, and wool and linen woven beaks.

This duality in the understanding of clothing's, and worn clothing's in particular, relationship to disease could operate within both miasmatic and contagionist models of disease etiology.[9] Nathaniel Hodges, appointed to the Royal College of Physicians based on his firsthand experience of the Great Plague, argued in his account of the plague that certain environments (miasmas) seemed conducive to plague, but the notion of discrete entities (seeds) as disease agents was also compelling. As Hodges wrote in *Loimologia*, articles of cloth and clothing became particularly suspect insofar as they crossed between theories: cotton, like silk and other fabrics, was considered "a strange Preserver of the pestilential Steams."[10]

This strangeness was inextricably bound to the movement of clothing between places and persons, as things that rubbed against life in all its forms. There were invisible particles, Hodges indicated, and yet they were not such as could necessarily be precisely located. Nonetheless, clothing provided an environment that either harbored or was itself implicated in these potential disease agents. The apothecary and medical writer John Quincy wrote a preface to the English language version published in 1720, reiterating the association between textiles and disease. His description at the same time revealed the way in which cloth in and of itself, once having been brushed against an ill body, could turn a disease particulate into a miasmatic environment and vice versa.

> The dissolved, and dispersed Particles may longer adhere to some inanimate Bodies than others, as to Woollen and Linen Cloaths, Papers, &c. and these Particles may, by the Steam of a living Body, or by the Means of any other Heat, be put into Motion, so as to breath out of those Lodgments, where they quietly resided, and obtain so much Liberty, and Action on all sides, as will carry them into the cutaneous Pores of any Persons within their Reach, and infect them; and on this Account a *Pestilence* may be brought from very distant Countries, lying a long Time in such Manner concealed, and then suddenly breaking out.

Animal fiber–based clothing is a linchpin of sorts between contagion and miasma, as concepts and disease agents alike.

> THE most common Manner of conveying and spreading a Contagion, observable in the preceding historical Collections, and which also is the Case of our present Apprehensions from Abroad, is by infected Persons, and Merchandize.... [E]ven Packs and Bails of Goods carry the poisonous *Miasmata* about with them; and from the Nature that we here suppose this Poison to be of, nothing is more likely to preserve it than animal Substances, as Hair, Wool, Leather, Skins, &c. because the very Manner of its Production, and the Nature of its Origin, seems to give it a greater Affinity with such Substances than any other, and to dispose it to rest therein until by Warmth, or any other Means of Dislodgement, it is put into Motion, and raised again into the ambient Air.[11]

Such was the *lively* nature of textile entities, that they could carry miasmata but also materialize and preserve the very poison itself; they rested and unrested between persons and things.

Consolidation of Clothes and Corpses

By the nineteenth century, bubonic and other forms of plague had taken a back seat as cholera broke out throughout Europe, beginning in 1830 and continuing throughout the century. Cholera was thought to originate from India, with which England was conducting brisk business in "Packs and Bails of Goods" quite precisely, brought into Europe by way of India's "noxious airs," according to miasma theory, or via "dangerous particles." Matters of dress were far from ignored, and the gravity of the situation was subject to both publicity and parody. A broadside produced in France—and subsequently translated and reprinted in Germany and England—makes light of efforts to ward off disease through "appropriate" dress. "He who would defend himself from the Contagious Cholera must be dressed in the following Manner," the extended caption begins. Exhibited is a pre-Victorian-era equivalent to a modern-day "hazmat" suit, though whether it is wholly or partially fantastical, one cannot be sure (fig. 3.1).

This cholera preventive costume involved "a Flannel six Yards long" wrapped around India rubber and pitch plaster, itself encasing the wearer. And beyond that, additional woolen layers: "flannel trowsers, thread Stockings dipped in vinegar, over which another pair of worsted, rubbed with camphire." The recipe for health through dress, the broadsheet implies, is both ludicrous and futile: "He must drag after him a cart containing fifteen yards of flannel" as well as a rather ungainly list and assortment of other items of wear. Ironically, it would seem that what the encumbered figure resembles most is none other than an itinerant peddler collecting and selling used clothing, a kind of archetypal ragman character of "Ralph Sly-boots" (as in fig. 1.1). And as the text ends, "by exactly following these directions you may be certain that the Cholera . . . will attack you the first." The aspect of the parodic here intertwines with and calls attention to an ambiguity in understanding: Was clothing ultimately a protective layer or a porous conduit for transmission? Were people just fooling themselves to even imagine otherwise?

The imperative of navigating the issue of clothing's relationship to cholera—its treatment, avoidance, and management—was not a laughing matter, however. Clothing's potential for the prevention and transmission of disease continued to allow the different theories of disease to coexist. On the one hand, many people were adherents of the miasma theory, holding variations on the belief that "corrupted" or "putrefacted" air was the root cause of—or alternatively the equivalent of—disease. Charles Dickens and Benjamin Disraeli are two whose accounts and thematizing of poverty in terms of rags coincided with a deep belief in miasmatic origins of disease, of its inhabitation of particular and polluted spaces, as—in the

Fig. 3.1 "Cholera Preventive Costume." Colored etching after J. Petzl, c. 1832. (Wellcome Library.)

case of Disraeli—Devilsdust's origins in a basement hovel replete with rags and bones. Henry Mayhew wrote extensively in the 1840s as a social commentator on old-clothes dealing and the neighborhoods frequented by rag dealers, describing their dangers in terms of "airs." Frederick Engels, the author of *The Condition of the Working Class in England*, was himself a miasmatist, seeing the environments themselves as polluted.[12]

Others, however, lamented that such discussions of miasmata were merely attempts to cloak what were in fact gaping lacunae in knowledge. As, in the years after 1832, cholera epidemics played a big role in the slow road toward the development and ultimate victory of germ theory, clothes and the rags that they became, and of which they might also have been made, played a role. It was during the major cholera epidemic of 1854 in London that John Snow famously discovered that the source of the outbreak could be traced to a single water pump, and from there to a piece of discarded fabric, a dirtied cloth diaper. Snow determined that the pump's contamination had resulted from a single piece of polluted cloth, the soiled diaper of a sick baby. When the mother had disposed of the used cloth, fecal matter ended up in the sewer system.[13] In an account of the 1832 cholera epidemic, *A History of the Cholera in Exeter*, published in 1849 by Thomas Shapter in the midst of another epidemic, water was still itself understood to be a barrier, a kind of natural quarantining system. But to Snow, it became clear that water was in fact the medium of transmission.[14]

The idea of these fomites—which at that time, as for Fracastoro, would have been defined as clothes, wooden objects, and jewelry, objects that, though not themselves corrupted, preserved living germ elements—made sense to London's denizens and Board of Health, and the city responded by shutting down the pump. People were beginning to have a stronger sense of this association between material things and disease and of clothing and its textile components as potentially lethal as well as protective. Nonetheless, Snow's theory of contagion was put aside. There are many explanations for this, one being that what would have been needed to transform the urban infrastructure and create a public health system was far too much for the city to deal with. Also, sterilizing different objects and discrete entities would have been extremely destructive to the way in which colonial relations were involved in the industrial system.

Even though miasma theory remained the preeminent understanding of disease throughout the early and mid-nineteenth century, the idea of discrete objects as fundamentally involved in disease transmission held increasing sway—especially in continental Europe, the countries of which by the middle of the nineteenth century were supplying vast quantities of rags to West Yorkshire while also setting up their own rag devils at home. Indeed, the shoddy industry was growing increasingly transnational at the very same time as these epidemic diseases ravaged. "There is hardly a country in Europe which does not contribute its quota of material to the shoddy manufacturers," a Scottish journalist commented in 1849. "Rags are brought from France, Germany, and in great quantities from Belgium." It is remarked that Denmark is "favorably looked upon by the tatter merchants, being fertile in morsels of clothing of fair quality" and cleanliness, in contrast to other nations, whose bales tended more toward "unmitigated masses of frowzy filth."[15] Fifteen years earlier, George Head had framed the philosophical quality of the shoddy in terms of disease potential, in what can be seen as an allusion to Carlyle's *Sartor Resartus* in referring to a "doctrine of the transmigration of coats" as necessarily binding into a simultaneously material and imaginative space "the blazing galaxy of a regal drawing room down to the cellars and lowest haunts of London, Germany, Poland, Portugal, &etc. as well as probably to other countries visited by the plague."[16]

In England and Europe, as well as in the United States, because apparel was a topic that played an active role in both theories, close attention to clothing—protective, on the one hand, and potentially contaminating, even toxic, on the other—and attitudes toward imported rags provide a lens through which to view interactions between these theories of disease as they operated alongside one another well into the later nineteenth century.

An image from an 1849 medical treatise, examined from the perspective of attention to clothing—its use, treatment, and potential dangers—is revealing. In the frontispiece to an English book by Thomas Innis called *The Skin, in Health and Disease*, skin disease offers a particularly interesting case where the theory of miasma comes into relationship with ideas of objects touching each other, with particular consequences for the status of used clothing and the shoddy cloth made from it (fig. 3.2). The image,

Fig. 3.2 Ringworm. Frontispiece for *The Skin in Health and Disease: A Concise Manual*, by Thomas Innis (London: Whittaker and Co., 1849). (Courtesy of Harvard Library.)

called "Ringworm," deals with the skin, the space where the living human being and the outside air or physical objects intermingle, with more or less porosity between them. Here, access to "unhealthy airs," resulting from improper clothing, and the presence of fomites come together. Depicted is a young girl of the upper class in the bloom of youth and the throes of skin disease: her head is covered with bloody pustules where hair and bonnet should be instead. One senses the value of self-care in her social context: clean blouse, clear peaches-and-cream complexion. That her head is marred, skin transformed into a grotesque mess, only makes the rest of her appear more pristine.

This image is significant to the book's unfolding: the purity of her shirt and the pillow on which she appears to be at ponderous and virginal repose accords with the value that Innis ascribes to proper clothing for maintenance of the health of the dermis and hence a proper conditioning of the self. What he proposes is a "regular system of cleanliness, a use of proper clothing, of which flannel next [to] the skin should form the principal item—and a prudent care as regards personal contact with persons and things to be suspected."[17] Health and vigor are the result of clothing chosen and deployed properly so as to maintain "the healthy action of the cutaneous functions" (45), with it being "the principle of difference in the conducting powers of bodies which guides us in giving a reasonable preference to certain preparations and fabrics over others" (58). The emphasis, as in Engels's *Condition of the Working Class*, is on the value of woolen flannel, specifically, "flannel next to the skin," a second skin.[18]

But which theory of disease is assumed here? The answer is both: contagion and miasmatic. The ringworm (*Porrigo scutulata*) described at the book's end, along with a proposed treatment with subcarbonate of soda, is characterized as "often be[ing] traced to contagion, consequent upon the circumstance of wearing the same cap," ringworm being something that can be carried from one person to another by objects, and in particular by articles of clothing and cloth. In this case, we see that the girl practically has a tonsure, as if she's been turned into a monk, except that the area that is marked out seems to be exactly where a girl might put her hat or some kind of head scarf. One imagines that she has either worn another's cap, whether friend's or stranger's, or perhaps a shoddy cap.

On the other hand, *The Skin, in Health and Disease* endorses the general belief among miasmists and others that certain clothes could ward off bad air. "Nature provides them [inhabitants of cold climates] on the spot with such substances as are best qualified; hence we find them clothed with the skins and furs of animals" (28). Like the oiled leather of plague times, wool would be a most effective boundary layer in climes damp, cold, and otherwise. It was simultaneously "armor-like" in certain ways (along the lines of the plague-protective costume), and porous at the same time, hence allowing for the circulation of air. But if wool was extensively praised in a wide array of contexts—contrasted with cotton for its durability—then shoddy was its miasmatic undoing.

This association between corrupted cloth and disfigured skin extended from ringworm to the larger health crises facing England in this period, especially as it pertained to efforts at prevention and remediation. A map from Thomas Shapter's classic on environment and disease, *The History of the Cholera in Exeter in 1832*, initially published in 1849, shows the twinned importance of the discard of human bodies and textile skins during the period of the first great cholera epidemic in England (fig. 3.3). Two clothing burial sites (marked "Clothes burnt and buried") and two human burial sites (marked "Cholera Burying Ground") are clearly labeled.

Shapter's *Cholera in Exeter* depicts through its images (engravings that were incorporated late in the production), as well as in its text, the extent to which concepts of contagion and miasma intermingled for people. Here again, the attention to clothing is telling. Even before cholera begins its journey as a kind of vector for the development and success of germ theory, cholera is figured, described, and visualized very much in terms of a public health crisis surrounding old clothes. Shapter's book describes the horrors of the cholera epidemic of 1832 in Exeter, outside of London (fig. 3.4). The image here is part of a panel that depicts how, at that time, a city under siege with this epidemic was dealing with it. What we are shown is clothing being removed from homes by a specific person who has been given this specific job. The appointed person is not dressed in the white hazmat suits of Ebola rescuers, or those who gathered the old clothing and bedding after the onslaught of cholera deaths in Haiti in 2010, or in anything even approximating the "preventive suit" we've seen. He is wearing ordinary clothes.

By August 1832, the Board of Health had decreed that city dwellers were obligated to give up both dead bodies and old clothing when asked, for proper treatment and potential disposal by the city. The person depicted here serves as one of the "inspectors" who "examined, condemned, valued and removed such of the clothes as they thought should be destroyed" (180). He went door to door during and after periods of

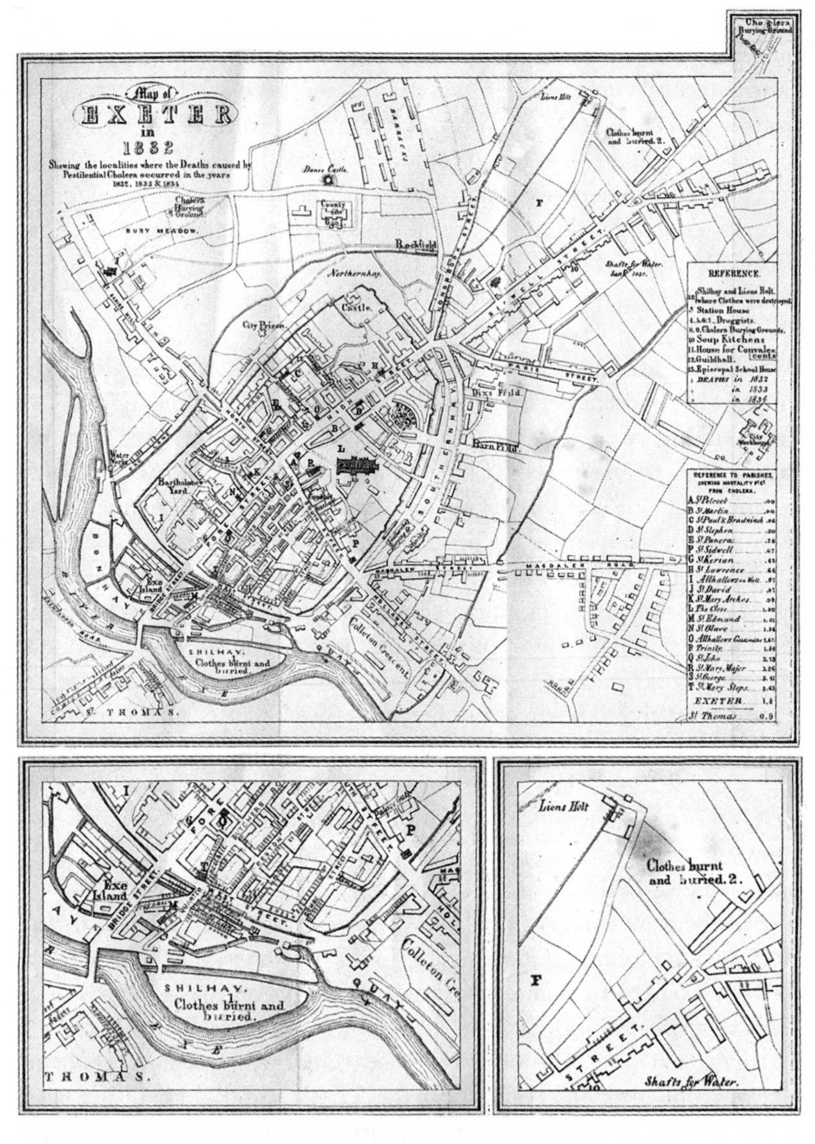

Fig. 3.3 Map from Thomas Shapter's *The History of Cholera in Exeter in 1832*, published in 1849 about the epidemic of 1832. (London: J. Churchill, 1849.) (Courtesy of Harvard Library.)

outbreak, simultaneously tallying the loss of human life and the value of household goods. Sheets, quilts, pillows, and clothes—a combination of bedding and wearing apparel packed in tightly under a canvas cover—would be taken away by cart. Specific directions for their examination, condemnation, valuation, and removal guided the process. Others would be sent for to pick up corpses. The air hung heavy with pervasive sorrow; "the removal of the clothes, though often evaded, was to a certain extent complied with, but frequently demurrings and threatening dissatisfaction accompanied the unwilling submission" (180).

Confiscated clothing had associations of sentiment, as well as value on the rag market. Could or should the collected clothes and bedding be cleaned? Was there a way to purify and then reuse it—to sell it for scrap or for shoddy? Much so-called "condemned clothing" would have been either burned or buried, but there were also attempts at "resurrection" and "remediation," cases where articles could be "reclaimed and distributed to those who need them" (182).

Purification through washing and "fumigations" were options and common throughout the city at various points, even up to the end, once the clothing graveyards had reached near capacity. "Some were washed in the fulling mills; others, after having been steeped in a leather dresser's lime pits for a day and night, were washed in the river, and rack dried" (182). And yet, even earlier on, concerns were aired and wrestled with about the appropriateness of "fumigation" alone in handling the clothes of the sick and dying.

The issue of how to deal with "sick" clothing became tied to managing the emotional needs of the inhabitants. A key for solving the clothing issue was preventing mass hysteria. Once the scale of the epidemic had become clear, burial sites were established for the clothing. As we've seen, a burial ground for human bodies and one for textile goods were established. Vehicles were identified for transporting each; textile and human hearses destined for cremation or entombment. The image on the left in figure 3.5 depicts denizens rinsing cholera victims' clothing in the river as part of an intended process of purification. Meanwhile, in the background, a horse-drawn carriage pulls what one assumes is a hearse. In the circumstances, it is not clear whether it is en route to a human or a textile burial. "Open caskets," as it were, were to be prohibited for clothes as

Fig. 3.4 Bridge Street and the old St. Edmund's Church. Textiles were examined, condemned, valued, and removed. From Thomas Shapter, *The Cholera in Exeter in 1832* (London: J. Churchill, 1849). (Courtesy of Harvard Library.)

Fig. 3.5 *Left*: Collecting clothes and corpses. *Right*: Burning clothes. From Thomas Shapter, *The Cholera in Exeter in 1832* (London: J. Churchill, 1849). (Courtesy of Harvard Library.)

well as for corpses, for it was found that simply seeing what under other circumstances would be a ragman's wagon heading through the street at such times was enough to cause an uproar. "A handcart, covered over with canvass, in order not to alarm passers-by, was chiefly used" (179). The covered cart thus would both conceal the contents and call attention to this: a visual death knell, though not death itself.[19]

Wool posed a special problem. Whereas cotton and linen—plant fibers—would come to have overtones of cleanliness, wool carried an aura of irremediable stink. It had an overdetermined quality, an associative doubling as a result of its animal origins, the coexistence of sick animals and people being an enduring possibility. This interconnectedness between human skin and sheepskin had long registered in ways both metaphorical and material, as in Isaac Claesz. van Swanenburg's Dutch Renaissance depiction of the virtual fusion of human and sheep integuments in his painting *The Removal of the Wool from the Skins and the Combing*.[20]

The etchings in Shapter's volume reveal this interconnectedness in a light specifically attuned to the disease element. Figure 3.5 (right) shows two men burning clothes at the Shilhay site, which would be later shut. Smoke is billowing, ballooning outward and upward, even encasing the lower half of one man's body. Behind them, we see the racks rising out of the dank air, where fulled woolens were stretched, this being a woolen town.[21] One can imagine how rags ground to dust became, in this context, the ultimate toxic miasma—devil's dust.[22]

Ultimately, the sites for burning clothing as well as clothing burial had to be moved because of the associated emotional intensity. The chairman of the Board of Health, in response to complaints, declared that "in order to obviate the feelings of excitement which have prevailed on the occasion, our Board of Health" had decided to "have buried the clothes of persons dying of cholera in a field out of observation" (183).[23]

Dealing with sick and dying people and the "sick" clothing left behind during an epidemic was only one part of the problem posed by epidemic diseases, however. The larger issue was how to prevent their spread in the first place. Here again, rags and their products played a role both in how the problem was posed and how solutions were formulated and attempted. As we have seen, shoddy, composed of rags from generally unknown and unknowable origins, could both exist as a material commodity and embody fears rooted in both miasmatic and contagion theories of disease. Shoddy itself was involved in efforts to grapple with the threats associated with it.

John Snow had, in a certain sense, solved the problem of how cholera spread in his analysis of the Broad Street pump issue. Nonetheless, many in England retained their belief in miasma as the ultimate root of disease. Across the English Channel, however, on the European continent, the contagion theory was much faster to catch on. Indeed, there doctors, public officials, and intellectuals were increasingly open to and accepting of fomite-based models of disease. The persistence of cholera precipitated more determined efforts to develop and systematize methods of sterilization. In 1851 people gathered in Paris to create standards for the quarantine of people and goods at the First International Sanitary Conference.[24] International delegates attempted to standardize regulations, such as there were, to prevent the spread of cholera during the movement of people and goods. The idea that goods might well transmit this kind of disease caused increasing amounts of suspicion and consternation, but the mechanism was unclear. And meanwhile, a great many commercial interests, from merchants to industrial capitalists, wanted to prevent burdensome restrictions on the movement of goods such as would occur if quarantines at points of passage and arrival became widespread.

Following the First International Sanitary Conference, other Sanitary Conferences were held. The British seemed relatively resistant to accepting concerns about clothing and commodities as themselves transmitting disease. Miasma theory—the belief that illnesses were carried by a cloud of smelly poisonous vapor created by decay—ended up being put to the test. As the most advanced account of how contagion occurs, germ theory, in a European context, was shown to be credible. However, in the United States, the 1850s were the years leading up to the Civil War, and there miasma theory still prevailed, and indeed had dominated conversations surrounding the outbreaks of cholera in the United States from the 1830s onward.[25] Certainly, ideas of infection and disease by the time the war broke out were not very well understood: hence all the deaths by infection and reapplication of used bandages that occurred during the war.

As the Sanitary Conferences continued to be held, however, disease was increasingly discussed not as something "in the air," but rather something carried in waste and on people's hands and even attached as textile skin to their bodies. But germ theory did not triumph right away, and advocates of public health struggled to prevent future epidemics. Rags played a role in both the conceptualization and ultimately the enactment of both theories. Rags equivocally connected persons and things, carrying traces of both human and object. Their very existence and the material and human conditions of their production rendered ambiguous the applicability of disinfection. As both discrete entities and, via shoddy, an amorphous substance, rags continued to constitute a middle ground between the two theories of disease transmission, an interesting figurative space, but also once again the physical space between competing ideas about disease. Shoddy, shredded-up rags, was simultaneously understood to be a fomite—a particulate of microscopic fibers, more or less corrupted, able to travel as vector from point A to point B—and to be somehow amorphous, a shapeless, shiftless (miasma-like) cloud.

The first two International Sanitary Conferences, both held in Paris at the beginning and end of the 1850s, would be remembered as key events in the emergence and institutionalization of the public health and "industrial hygiene" movements. The Second International Sanitary Conference met eight years after the first, in 1859, on the eve of the outbreak of the Civil War in the United States. Prior to the meeting, despite the success of Snow's proposed solution to the 1854 epidemic and the continuance and even worsening of cholera epidemics in England in the preceding several years, skepticism of his analysis and proposed solution had persisted. However, in this decade, new investigations conducted on other diseases that had particular consequences for both the general well-being of individuals and of national industry on a broader level had begun to shift viewpoints, even in England.

During the 1850s, anthrax, a disease with a profound effect on animal husbandry, broke out with increasing numbers of cases in England, as well as throughout continental Europe. Louis Pasteur, in France, had taken up the issue of anthrax in the mid-1850s, where—as in England—its impact was felt in the woolen industry, both in terms of the labor force and raw-material production. Anthrax decimated sheep populations, and the disease was ultimately transmitted to the other animals and to the humans in their midst—from sheep to shepherds to sheep shearers to workers in the fulling mills. In the process of developing a vaccine for anthrax—the effectiveness of which Pasteur would eventually demonstrate throughout the French countryside—Pasteur identified the cause of anthrax. It was a microorganism, carried from one host to another through a vector: generally something palpable and material that could be transmitted, in fact, on such an object as the clothing worn by the shepherd or by the sheep's wool and, indeed, the very kind of commodity whose production motivated the whole set of intimate interactions between man and animal (fig. 3.6). This research on anthrax helped make the British a bit more open to the idea of there being microscopic entities that inhabited such likely things as clothing and wool. British resistance further crumbled during smallpox epidemics that struck over the next several years. Though the discussion at the Sanitary Conferences remained focused on cholera prevention, discoveries in these other domains were mobilized to argue for a contagion model of cholera as well.

Concomitantly, suspicion of rags, but especially of imported rags, began to increase, as the idea of rags being vectors (fomites) grew more and more resonant. A lengthy report issued in England prior to the Third International Sanitary Conference was called "The Question of Injury from Infected Rags." The report was produced under the authority of the chief medical officer, John Simon. Its author was Dr. John Syer Bristowe, a respected member of the Royal College of Physicians, who twenty years

Fig. 3.6 "Louis Pasteur Inoculating a Sheep against Anthrax." Engraving from Charles Edouard Chamberland, *Le charbon et la vaccination charbonneuse* (Paris: B. Tignol, 1883), p. 304. (Courtesy of Harvard Library.)

later would be elected a fellow of the Royal Society. A practicing pathologist at St. Thomas Hospital and curator of its pathology museum, Bristowe was somebody with a clear interest in working with objects that were both material and connected to the body.[26]

In the 1860s, studies produced in England and by delegates to these International Sanitary Conferences were read with great interest in both the United States and in Britain. In 1866 a series of reports on rags were issued. In England, *Public Health: Eighth Report of the Medical Officer of the Privy Council* was put out concerning the issue of rags in the context of epidemics. Similar articles appeared elsewhere in Europe and North America, with the issue of sterilization also entering into the conversation.

In the period after the Civil War in the United States, the rag industry expanded tremendously: in terms of both the volume of its commercial output and the reach of its negative reputation as all but inherently corrupt and corrupting. There, the National Association of Woolen Manufacturers (NAWM) was formed precisely at the moment when shoddy's reputation was nearing its nadir and as rags' connection to disease was a topic of international discussion. Founded in 1864, at the end of the Civil War, it devoted much of the first issue of its journal to coverage of the shoddy industry in general and to the issue of contagion in rags in particular. Many veterans suffered from gangrene and lifelong diseases as a result of their service. Shoddy had come to both concretize and abstractly represent the corruption and lack of patriotism on the home front during the war. The issue of contagion in general, and of rags and contagion in particular, was something that the NAWM was interested in finding a way to dismiss. It did so by casting doubt on whether rags actually transmit diseases and on the whole theory of contagion. The burden of proof, they claimed, was on those who claimed otherwise.

The first publication by the NAWM was a transcript of "The Fleece and the Loom," the plenary address given by John L. Hayes at the first annual meeting of the organization in Philadelphia.[27] In that talk, Hayes had included a discussion of international concern over the import and export of rags. Later, in the first issue of that organization's trade journal, the rag question was highlighted substantially. Drawing on material from the *Public Health: Eighth Report of the Medical Officer of the Privy Council*, the NAWM called attention to Russia, Egypt, Turkey, Europe,

India, and China, all countries from which shipments of rags frequently originated. Without specifically referencing the British 1865 report, which by this point had been reproduced elsewhere as well, an article described the shoddy industry in detail:

> These woolen rags are collected, packed in bales, and are imported from various areas . . . from districts where plague, fever, smallpox, and all sorts of loathsome skin diseases extensively prevail. The bales are opened and the rags are sorted by human fingers before being placed in machines which break up, tear, separate, and cleanse the fiber for manufacturing uses.[28]

The author then lists various categories of illness—the "loathsome" skin diseases such as ringworm, plague, and smallpox, as well as cholera—admitting their presence in the various countries of origin of rags. The contention he makes, however, is this: "According to the evidence we obtain, no disease has ever broken out among the persons who so manipulate these old woolen rags. Although in many of the countries in which they are collected they are believed to be peculiarly plague-bearing materials." He simply denies that rags could be responsible for transmitting disease to the people who receive bales at the docks, sort them at the mill, or shred them in the devil, because "the lapse of time and space in collecting, storing and transmitting these rags and also the destruction of any such poisons by friction or otherwise, whether possible or actual otherwise must be taken into account." Stepping back from the overseas public health and medical reports, *The Fleece and the Loom* concludes: "The whole of the facts deserves, however, the serious attention of those persons who insist that the power of communing disease is contained in a dangerous manner by woolen goods which have been worn by persons suffering from contagious diseases."[29]

In other words, it may or may not be the case that rags or secondhand clothes can transmit diseases, or even that such things as germs exist at all. Regardless, however, the burden of proof on the power of communicating disease is on those who make that charge alone. This was a point of view shared by many others on both sides of the Atlantic. An 1866 report regarding the spread of contagious diseases presented smallpox, based on its outbreak in and around several rag and shoddy companies in Northern England in the 1850s, as being the most likely to be linked to the collecting, buying, selling, treating, and transporting of rags. In that publication, officially a "report on inquiries into whether the RAG TRADE is of influence in SPREADING INFECTIONS of DISEASE," and a follow-up to the manuscript *On Infection by Rags and Paper Works* of the previous year, Bristowe distinguished between domestically and foreign-obtained rags by the "powerful compression" involved in the transport of the latter (foreign rags) and a lack of standardized sanitary precautions in the former (domestic rags); and yet he describes the passage through many hands of varying degrees of cleanliness and healthfulness in either case. Bristowe pondered: "How is it (it may be asked), if infected rags are capable of carrying contagion, that diseases are not more frequently conveyed by rags than seems to be the case?"[30] Proof of danger, and hence of mandatory quarantine combined with sterilization, would need to include ruling out these factors as adequate to prevent the transmission of disease.[31] Without proof showing otherwise, time and distance and friction should be assumed to render fear needless.

As the hygiene and sanitation movements gained momentum later in the nineteenth century, however, theories of contagion dominated meetings concerned with public health. In the 1870s and the 1880s, debates about germ theory coincided with debates about both the necessity and the means for disinfection as smallpox epidemics increasingly became a major issue. By then, Robert Koch was beginning his work on what would become known as the first postulates of microbiology. He spoke at the Fourth International Sanitary Conference held in Vienna and then in 1876 presented his discovery of the anthrax bacterium by means of its isolation. Soon after, he would similarly identify the cholera bacillus and then the bacteria causing that great menace, tuberculosis.[32]

Disinfection and Its Discontents

Ideas about sterilization that had been considered but marginalized in previous decades now took center stage. Disinfection and asepsis made more sense once disease agents could be localized. England's Lord Lister had, in the 1860s, developed what was at the time the first major approach to aseptic surgery: carbolic acid.[33] In hopes of avoiding the kind of mass casualties resulting from the surgeries during and after the American Civil War, surgeons began using both the acid and the apparatus that he had invented, even as ideas of infection and contagion continued to be dismissed.

As early as 1865, connections between rags as articles of trade and potential vectors of disease had been laid out clearly in the medical literature, as by the Medical Officer's Report that summarized Bristowe's findings described in the previous section.[34] With cholera still a key concern, by the time the Fifth International Sanitary Conference met in 1881 in Washington, DC,[35] however, germ theory had gained ground, and prevention had become the key issue. Public health officials drew on developments by Koch and various articles in British and European medical journals about rags while legislators, health advocates, and textile industrialists debated current and future laws and regulations surrounding the importation of rags and rag-containing articles of merchandise into the United States.[36]

Once germs were seen as real, how could they be removed from desired commodities or the raw materials from which they came, as well as from all those entities that existed somewhere between the former and the latter: from rags, used clothes, and shoddy? Mattresses posed a particular point of contention as they were so often filled with shoddy, so hard to get inside of without destroying, and spent so many hours pressed against a sleeper's body.[37] One mattress manufacturer working with only "new materials" presented the American Public Health Association with concerns about—and samples of—the "vile stuff" being thus squirreled away, largely by immigrant workers in New York and other urban bedding manufacturing centers, and yet marketed under the "innocent title of wool bedding"[38] (fig. 3.7).

The prominent theme was quarantine. As the bacterial origins of disease and the association of particular microorganisms with illnesses (cholera, anthrax, smallpox, etc.) became clear, rags and the importation of rags to make shoddy became increasingly entwined with issues concerning both the importation of foreign goods and with immigration of people from foreign lands. People were bringing rags into the country by carrying them on their own bodies or in layers and layers of clothing wrapped

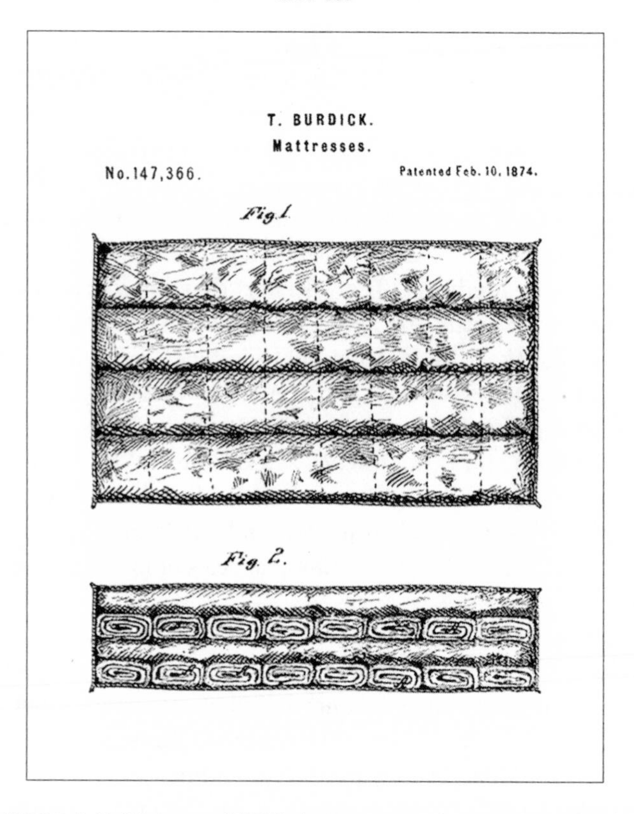

Fig. 3.7 T. Burdick's "Mattresses" (1874) shows one way of incorporating shredded-up rags into mattresses to be marketable under the innocent title "wool bedding." (United States Government Patent Office.)

on their bodies so they did not have to pay for a suitcase, and people were coming off boats with their baskets overflowing with fabric and old pieces of clothes. The clothing that people who were deemed undesirable either wore or collected and sold for shoddy elicited fear and suspicion as to its origin and content.

A series of bans was placed on the importation of foreign rags almost simultaneously with the announcement in 1884 of the identification of the cholera bacillus.[39] Moist cloth of any kind, it was posited, could carry the bacteria; given the lack of understanding of the effects (whether negligible or decisive) of quarantine, prohibitions seemed wise in many instances.

Fig. 3.8 Friedrich Graetz, "The Kind of 'Assisted Emigrant' We Can Not Afford to Admit," *Puck*, July 18, 1883, centerfold. (Courtesy of Harvard Library.)

In political cartoons of immigrants and disease arriving in American ports, on the one hand, and pointed social documentary photographs of immigrants and their rags once on solid ground, on the other, immigrants and rags fused together in the popular imagination, as can be seen in representations articulating both the promise and horror of such vital intermingling of persons and things, and the intermixing of categories and forms inherent in each. The cartoon in figure 3.8 appeared in the American humor magazine *Puck* a year before Koch's announcement of the cholera bacillus. Based on a drawing by famed German American artist Friedrich Graetz, it appeared as the centerfold of the July 18, 1883, issue. Captioned "The Kind of 'Assisted Emigrant' We Can Not Afford to Admit," it shows a ship waving a British flag coming to New York City. A skeletal figure seems to represent death itself; the figure carries a scythe with the band on its belt bearing the identifying name *Cholera*. A raggedy blanket on his lap, a tattered sheet over his shoulders, he is shrouded in woolen attire. The red woolen vest and red wool fez both evoke the Asiatic countries from which most cholera was thought to originate before reaching Russia in 1817.[40] Far below is a tiny American boat full of workers from the Board of Health; *too little, too late* is one's immediate impression. Disinfection and quarantine policies had become a subject of great debate in the preceding decade and a half; and yet it was only in 1881 that the United States had begun to participate in the International Sanitary Conferences. George Sternberg, head of the New York Board of Health, mobilized a response upon cholera's arrival in the United States.[41] But the Board of Health rowboat here is no match for this gigantic ship from abroad, and it is not only the rowboat but also the various characters lined up on the shore that come across as weak.

The ship is arriving at Castle Garden, the fort-like first immigration depot in the United States, initially administered by the federal government, though by this point under management by the state of New York. Empty bottles labeled "carbolic acid," "thyrol," and "chloride of lime"—all chemical compounds thought to have aseptic or disinfecting properties—are positioned like military cannons around the wall of Castle Garden. Alas, they are cannons with no cannonballs: all show, with no actual defensive capacity. Behind these impotent characters is a sign welcoming the arrival of immigrants. One gathers that this is a commercial ship bringing in goods that had gone from India to Britain and then from Britain to the United States. The woolen felted cap has ribbons of hair-like fibers at the top. And beneath it is a vast, almost obscene amount of long, thick black hair. Its lushness is uncanny, as if it continues to grow and expand, even as the figure itself shrivels into only rags and tatters of skin and bone.

The relationships between rags and people, between living and nonliving things, thus come together in the depiction of international trade, immigration, and the flow of capital. In the title, "assisted emigrant" refers specifically to a series of policies under which certain countries (Britain being one example) paid to send away various undesirable elements of their population.[42] An individual would be given money, or a trip would be heavily subsidized, so as to "offload" these undesirable human elements, dumping them on foreign shores. In the Graetz centerfold, the assisted emigrant is a zombie-like figure, a human both living and dead, who carries in the feared disease. That Death is shown with emaciated skin, a rather unusual depiction thereof, illustrates the in-between nature of germs (living or dead?), shoddy, and clothing more generally (between self and environment, lively and dead). The conduit of disease would seem to be the manifold rags hanging off his raggedy skin, and the image points to the problematic nature of rags—the unknowable environments from which they hailed, the bodies they encumbered, and the people, places, and things that would be displaced or destroyed in the wake of their arrival.

Over the course of the decade, as the number of immigrants increased dramatically, individual states and agencies produced reports about the dangers of rags, as well as possible remedies for those dangers. Popular medical literature had pinpointed rags (rightly or wrongly) and other used textiles as fomites par excellence; "the most important fomites are bed-clothes, bedding, woollen garments, carpets, curtains, letters etc.," as *Quain's Dictionary of Medicine* had it.[43] George Sternberg, who later became the US Surgeon General, oversaw and represented the Committee on the Disinfection of Rags at the American Public Health Association while managing the New York Board of Health. In 1886 the latter organization issued a report on the topic (authored by chief customs officer W. H. Smith), covering a four-month period (June–September 1886) on the occasion of the board's establishment. By far the largest topic and the most extensively treated subject (at seventy pages) was rags, the discussion of rags being followed by much shorter sections on food, milk, and drugs. Of critical importance were both the means by which rags were imported into the country and the way they were handled by sorters, shredders, and ultimately the mills that would create new goods from them (fig. 3.9).

Interrelated and yet distinct categories of questions were pertinent, different ones coming to the fore at different times and in relation to differently "packaged" rags. Could, or, in a specific instance, *did* rags coming into the United States carry infectious disease? If so, how was risk in any particular case to be evaluated, and what measures would be most effective in that particular case? Quarantine? Disinfection? Outright prohibition? Various constituencies—representatives from the US Treasury, the corrupt rag rings of Brooklyn, and the Surgeon General among others—debated these issues.[44] Two general areas of concern, as well as individual sub-queries derived from them, depended on the category of rag: whether they were individual articles or compressed bales.

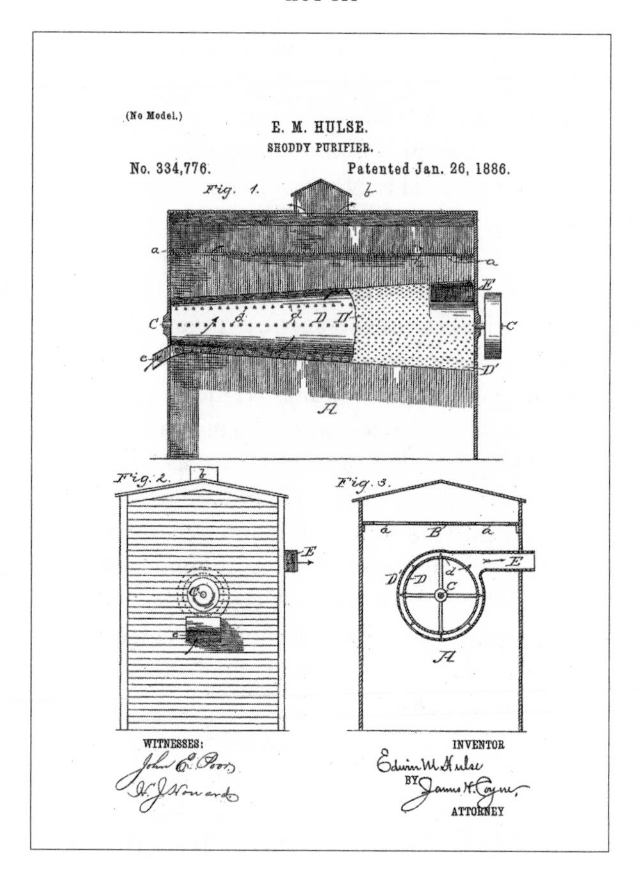

Fig. 3.9 E. M. Hulse's "Shoddy Purifier" (1886). Inventors claimed that such a device would evacuate whatever disease agents might still be present after shredding, as well as getting rid of additional otherwise unwanted bits and pieces, dust, and detritus. (United States Government Patent Office.)

Strategies for disinfecting one type might not work in the case of the other type. For an adherent of the contagion theory, an individual piece of clothing is a prime, even an exemplary, fomite and ripe for interpretation as such. A bale of rags, however, presents something quite different; the clothes' radical compression creates a blocky kind of textile environment in which, potentially, many disease agents could be concealed and prosper at an imperceptible damp core. Regulations could differ regarding the entry of each: individual rags or those carried by an immigrant in a suitcase or basket would be confiscated and burned, whereas compressed bales might be let through without need for quarantine or disinfection.

And yet the free-floating articles of old clothes tended to appear more ominous. When a rag has been carried on the back of an impoverished and potentially sick immigrant, the rag itself is like a remnant or a trace or a shadow of the person. When people are bringing in rags, the rags are in a sense part of the people, whereas in a bale, they are more like a skin that has been shed and left behind, gathered together as an amorphous mass: more like coal than corpse.

The British wanted to avoid barriers to the rag trade, given the scale of the shoddy industry in England and the magnitude of trade more generally between the UK and its colonies in the Far East (especially India and Egypt). Both India and Egypt "produced" rags and were also rife with illness, although there were relatively fewer immigrants coming to Europe from these places at that time, and the "piecemeal" rags tended to be coming from Russia and eastern Europe. So quarantine, some method of disinfection, or the exclusion of people and objects from infested areas appeared to be necessary. The sterilization approach was complicated, however, by the difficulty or near-logistical impossibility of sterilizing rags at an industrial (bale-sized) scale. Individual articles or small piles of clothing could be placed in solutions of lime or carbolic acid.[45] Other options for sterilization included boiling, steam heat, dry heat, and chemical treatment. Loose heaps, meanwhile, could be "baked" in carbonizing ovens (fig. 3.10). For the bales, denominated "ordinary articles of merchandise," the situation was more complicated.[46] Some wondered if there could be a way to modify chemical protocols by, for example, an "injection method" that ideally would facilitate deep penetration into the bale. A steam-heat method was plausible but would leave the center of the bale thoroughly damp and with no means to dry out, on its way to becoming wool both rotten and infectious.

However, the cost of anything else—especially anything that involved immediately breaking apart the bales—was seen as prohibitive. Most disinfection methods from this period were based on steam. However, the economics of using even these made their implementation difficult, especially in the case of bulk rags for shoddy, where the margins were slim

Fig. 3.10 Baking room with carbonizing oven for rags. Rags are seen on tables. The temperature was kept at 200 degrees Fahrenheit for about eight hours. Henry G. Kittredge, "Shoddy: Or the History of a Woolen Rag," *Technology Quarterly* 19, no. 2 (1906): 74. (Courtesy of Harvard Library.)

already. Consequently—as in the instance described in the court hearings excerpted at the outset of this act—there was an incentive to admit these giant bales without any sterilization or quarantine, even when they came from infected, infested, cholera-stricken places, and even as both the people and the individual rags coming from these places were quarantined, rags destroyed by burning, or immigrants turned back. It wasn't until the early 1890s that cholera outbreaks in England finally led to concrete discussions in the House of Commons about the necessity of banning the importation of rags.[47] And in the United States, there was substantial disagreement about the viability of either quarantine or disinfection of any kind. A preeminent specialist in matters of public health and disinfection was of the opinion that rags, while possibly not a cholera risk, were definitely a conduit for smallpox, but that detention in quarantine would not prevent these from carrying infection; and as there are difficulties in their disinfection, it only remains open to prohibit their importation, and to destroy them, if, nevertheless, they arrive—fire in his opinion being the only true disinfectant.[48]

Like the British, regulators in the United States distinguished categorically between singular and bulk rags. And like them, they favored the importation of bulk bales and discriminated against rags borne by individuals from the same areas. Amidst rumors of Asiatic cholera's arrival in England, US medical experts in the 1890s distinguished between clothing worn by immigrants or discarded bandages or hospital linens and so-called "rags of merchandise." An article, "Cholera in Rags," reported that "the whole matter was gone into at the Dresden conference, and that it was then found impossible to lay hands on a single case in which infection could be traced to rags imported in compressed bales as ordinary articles of merchandise. It is quite otherwise in regard to the loose soiled linens of travelers from infected districts."[49]

Restrictions placed on the bales, as opposed to individual rags, would be useless, the argument went, especially if they were placed only when an outbreak happened to be under way at the shipment's point of origin, given the gaps of space and time between outbreak, goods, and point of arrival.[50] Banning based on the point of origin also would be complicated; a bale of rags might come from Egypt, land at Holland, and continue on to London. Consequently, prohibiting the importation of rags would have to be an all-or-nothing affair. And those who made the argument assumed that no one wanted to ban them entirely.

> It is worth bearing in mind that it is believed that the rags of merchandise are sometimes years in reaching their destination at the shoddy mill, passing as they do through the hands of collectors and dealers of various degree. Not only then is it quite unproven that such rags have ever given rise to cholera, but a consistent attempt to stop the introduction of infection by their means would demand an almost permanent prohibition on their importation.[51]

Yet the epidemics continued, and the linked phenomena of the importation of rags and the arrival of immigrants from distressed areas of the globe continued to be seen as important vectors for the transmission of disease. At the Vienna Sanitary Conference, held in January 1892, the first international quarantine laws were set down. Over the following six months, dispute and debate necessitated modifications, and all parties signed off on the restrictions by late summer. On cholera, which was again the focus, instructions stated that its transmission was effected principally by the digestions and vomited matter and, consequently, by linen and clothing. With this in mind, the rule that ultimately went into effect by the year's end pertained to the quarantine and possible exclusion of both human bodies and textiles. Both people who were visibly sick and old clothing, rags, and bed linens (wool blankets, comforters, and mattresses as well as various cotton items) were to be isolated, disinfected, prohibited from entering, or some combination of all three. By contrast, non-textile items of commerce—whether new or old, such as old letters, newspapers, books, or ceramic ware—were to be free of all restrictions.

Just as the United States was signing on to this first "health protection" treaty by regulating trade in old clothing, it was also updating its immigration legislation. The first American general immigration law was just over a decade old at this point, having been passed in 1882.[52] These initial regulations excluded criminals, the insane, and others unable to support themselves from immigration.[53] The 1891 update expanded these exclusion criteria dramatically to include those who were ill, where sickness would be established by either visible evidence or other sources of "strong suspicion." Until this point, contagion and disease had not really been specified clearly as qualities to be used against somebody, at least at a federal level; now people could be blocked from entering the United States on the basis of appearing sick or being suspected of having an illness. Whole classes of people from places where there was or might be a lot of disease, though not officially banned, would be subject to a new level of analysis.

In the 1892 chromolithograph "They Come Arm in Arm" (fig. 3.11), featured in a very elaborate back cover of the September issue of the American humor magazine *Judge*, Uncle Sam, already a well-recognized personification of both government regulation and nationalist sentiment, peeks out from behind the walls furtively as three figures—immigrants of various kinds—arrive at the gates.[54] Even if one of the travelers is named

"Asiatic," he, too, comes directly from Europe. Indeed, there is a doubling down on the European origins of the situation at hand. In the distance is a sign marking the boat's point of origin: "Europe." And the boat's flag carries only the words "From Europe" in large block letters. One immigrant looks typically Irish, another Jewish. The latter is covered in rags, has rags streaming out of his pockets, and is carrying what looks like a bundle of rags stuffed into a pillowcase. The Irishman's accoutrements are in only slightly less disarray. Connecting the two "live" arrivals is a death figure wearing a ragged cloak and a prominent name tag: "Asiatic Cholera." The rags tie the three figures together—visually, spatially, and (via the sharing of microorganisms through textiles and skin contact) physiologically.

The federal government had in recent years been so concerned about ineffective (to say nothing of corrupt) immigration controls that until 1890 it had taken over much of the management of the process from the states. The Ellis Island Immigration Station had opened on the first day

Fig. 3.11 "They Come Arm in Arm: American seaports must close their gates to all three," *Judge*, September 1892, back cover. (Courtesy of Harvard Library.)

of that year, replacing the Castle Garden Emigration Center, the immigration depot shown in the 1883 Graetz illustration. The opening of the huge structure on an island recently expanded with leftover city subway ballast had been followed by almost half a million human arrivals by the time the image was published.

The differences between the two images are thought provoking. Rather than being figured as skin and bones, as in Graetz's 1883 image of the "Assisted Emigrant" (fig. 3.8), this 1892 image of illness has no skin attached at all: it is rags and bones alone. He is a "rags and bone" man, the name by which old clothes peddlers would often have been referred to at that time. Rather than carrying a scythe (as in the Graetz image and in depictions of Death generally), this figure instead grasps the two raggedy men; the Jewish and Irish immigrants have become his sidekicks and weapons (vessels, carriers). They are his technologies of choice: human scythes. And whereas shock seems to have made the Irishman drop his luggage before reaching the gate (perhaps he will turn back, given the inauspicious welcome), the other arrival, closest to the viewer, seems only to clutch his bag of rags more firmly, bearing on his face a look of grim defiance.

This is someone who, like the Death figure, is destined to be a "rags and bone" man himself. The slightly turban-like Russian-style wool hat is a tip-off, even beyond the caricatured typically "Jewish" features. Though he is not dead (yet), illness may be in his future soon enough. At the very least, once (or if) he is let in, always a wanderer, he will be carrying old clothes between neighborhoods and hence be a possible vector of transmission for illness from, let us say, the poor infected with tuberculosis in Lower Manhattan to the rich uptown. He may at present be carrying rags from overseas that will infect future handlers and wearers. But the implication is that once he arrives, his occupation—and perhaps the very essence of his being—will be the transmission of disease from one wearer of a garment to another; he seems to exude rags in all directions. Cholera, meanwhile, is literally held by and holding old clothing, dirty clothing, soiled hands: immigrants hold dirty clothes while also holding hands with Death, rubbing shoulders against the sickly blankets with which the walking skeleton is covered.

Rags here stand in for undesirable commodities and undesirable persons; they are bound together with disease, in this image as in the new legislation with which its appearance coincided. The question of the arrival of foreign rags on native soil is tied through this figure to the manner by which carriers of such rags may lead to the pernicious arrival of new international admixtures, both textile and ethnic, commercial as well as interpersonal.

The immigrant marked as an eastern European Jew is of course a stock character whose association with rags endures in this period.[55] In *The Familiar Man Who Buys Old Clothes*, a photograph made by the American Press Association, the name of the subject is not identified. The photo depicts, in essence, a type specimen, a synecdoche—a bag of rags that stands in for all bags of rags and an immigrant who stands in for all immigrants (fig. 3.12). Like the bag of rags, the "familiar man" is of his particular moment, as well as playing off of a centuries-old stereotype. Considered in light of the moment and medium of its production, it also reveals a simultaneous dustiness (the dirty streets and satchel) and inscrutability of such ethnic environments, and the corpse-like aspect of the old clothes. Who wore those clothes? Were they sick or healthy, foreign- or native-born? And how could this man, who himself is alien, become something in society that is familiar?

The association of immigrants and their rags is especially clear in the depictions of women, who come to seem, at times, virtually sculpted from rags. In photographs made of their passage into the country and as inhabitants of ethnic enclaves, these images proliferate. Whereas men tend be shown carrying rags in satchels, barrows, or baskets, women are often encumbered, rendered almost sculptural by the addition of the woolen layers. Women would often cross the Atlantic wearing multiple sets of clothes, article upon article, their own as well as those of other members of their family, for the passage. The obscuring of faces and bodily contours is not for modesty's sake, but rather for maximizing baggage (carried as "worn" as in the French *porter*). The photograph *A Tenement Gleaner*, by social documentarian Lewis Hine, provides a sense of elision between women's bodies and articles of secondhand textile commerce (fig. 3.13). A woman is shown carrying an enormous bag of gleaned rags; she is an interesting figure because she is even more associated with rags and with clothing and cloth than the "familiar man." We see clearly only her thin

Fig. 3.12 *The Familiar Man Who Buys Old Clothes*, American Press Association, no. 839. (Wallach Division of Art, Prints, and Photographs: Picture Collection, The New York Public Library.)

hands and wrists as she is otherwise covered in several layers of cloth, multiple skirts, and a black scarf.

The Intimate Materiality of the Unknowable

Shoddy, as commodity, as allusion to the process of its manufacture, thus was closely knit in with a range of visual and material connections. In the later decades of the nineteenth century, the shoddy industry was thriving,

Fig. 3.13 Lewis W. Hine, *A tenement gleaner, New York City*, c. 1908. (Wallach Division of Art, Prints, and Photographs: Picture Collection, The New York Public Library.)

even as the business in paper manufacture from cotton and linen rags was undergoing rapid changes as a result of the emergent wood-pulp paper process.[56] Associations between immigrants, old clothes, and disease and its unseen manifestations being pervasive, developments in disinfection technologies were both hailed and rejected.

The simultaneously anonymizing and deeply personal aspect of shoddy—that it carried the traces (whether pathological or simply amoral) of individuals, on the one hand, but existed as a mass of homogenized raw material, on the other—conflated the intimate and the unknowable. As a "mound of the unknown," it became a commodity, processed and re-formed in such a way as to wind up in a new set of intimate relationships with the human body—blankets, mattresses, army uniforms, ladies' jackets. So that by the century's end, even though the issue of disease—its origins and how sterilization and disinfection could be achieved—was better understood, shoddy maintained its peculiarly evocative nature as to issues of infection and otherness. Only it had moved, now, from the realm of the somatic to that of the moral—from physical disease to fear of ethnic admixture and its effects on social structure and commercial culture alike.

As both material and process, shoddy could be conceived of as a metaphor for the "melting pot," itself an image fleshed out into story and metaphor for the purpose of giving voice to the positive contributions of immigrants arriving in the United States. Israel Zangwill's *The Melting Pot: The Great American Drama*, first performed in 1908, is represented in a striking 1916 promotional poster (fig. 3.14). The graphic shows the Statue of Liberty with people arriving from far and wide, their paths passing through the radiating rays of sun before converging into the patriotically branded admixture. Ultimately they become the human elements—the weft and warp perhaps—of an American admixture to be celebrated (as has largely been understood to have been Zangwill's own intention), or denigrated and despised by nativists.[57] This concept of the "melting pot" crucible bears no small connection to the figure of the rag grinder as a positive social as well as technological apparatus. Shoddy was promoted as civilized, productive as enterprise and commerce, and forward-looking in terms of both industry and society. And even its less savory aspects could be redeemed, "purified" as in the Shoddy Purifier in figure 3.9. Indeed, like the melting pot conceived of a decade later in the realm of theater that eventually entered into common parlance, shoddy in and of itself could be America's particular kind of promise and product.

Such images and text existed in tandem with those of imported rags and shoddy as threats to the fabric of American life.[58] A lavishly illustrated and elegantly bound volume titled *The Shoddy Industry vs. the Virgin Wool Industry*, published in 1919, framed opposition to the shoddy trade in ideological terms. The National Sheep and Wool Bureau of America produced the book as part of a major campaign to tear down the shoddy industry, and to do so specifically by forcing the passage of laws that would

Fig. 3.14 "America is God's Crucible, the great Melting-Pot where all the races of Europe are melting and re-forming!" Text from *The Melting Pot* (1908), a groundbreaking play by Israel Zwangill, poster design, 1916. (University of Iowa Special Collections.)

require labeling shoddy products clearly, rather than allowing them to be, in the bureau's terminology "camouflaged," thus allowing them to "pass" as "pure" or "virgin" wool.[59] Shoddy, it was argued, undercut sheep husbandry. Direct associations are made between the blending of various derivatives of wool (e.g., shoddy, mungo, adulterated wool, "virgin" wool), and the blending of immigrants (the kinds that tend to collect and be associated with rags and dirt) with hale and hearty "old stock" Americans.

"Moral Blight, Soldiers Come Home," an illustration from *The Shoddy Industry vs. the Virgin Wool Industry* (fig. 3.15, top), depicts soldiers returning from World War I. Three boats enter the harbor, carrying excited veterans including the maimed and battle-scarred, one of whom we see in a wheelchair huddled up under a ragged and hastily assembled patchwork coverlet. They look up with a mixture of joy and relief, facing toward the sunshine radiating out from the crown of the Statue of Liberty, perhaps blissfully unaware of the dangers posed by rags and the "melting pot" of the shoddy industry. The accompanying text dramatically warns:

> Among the boys on these transports who've been through hell in defense of truth and justice are many who, because of shattered health and nerves and crippled bodies, have paid an even greater price than would have been the loss of life itself. The practice of selling shoddy in fabrics and clothes without letting its presence be known is a moral blight and an economic evil that threatens our institutions. The truth and justice in defense of which the war was fought and won will continue to be set at naught, so long as the practice of camouflaging shoddy is permitted.

The suggested cause of both moral and economic decrepitude is both shoddy and the makers of shoddy—the rag collectors, sorters, and shredders. *The Shoddy Industry vs. the Virgin Wool Industry* takes pains to personify this enemy as a single immigrant ragpicker. Another illustration called "An Amazing Discovery" describes its subject as "one of the vast army of rag gatherers" (fig. 3.15, bottom). It is precisely such a "vast army" that we have seen flowing into New York City Harbor, guided by the Statue of Liberty, in the "Melting Pot" poster in figure 3.14.

The shoddy industry itself had, for decades, been combating this kind of framing of its product in terms of depraved material, ethnic, and moral origins, cleaning up its reputation and eventually even developing new names for the entity (*reworked wool* and *recycled material* among them).[60] In this regard, consider the images and text from a popular magazine article from 1896 in *Scientific American* showing the process of shredding up rags to make shoddy clothing (fig. 3.16). It is a highly illustrated

Fig. 3.15 *Top:* "Moral Blight, Soldiers Come Home—The practice of selling shoddy in fabrics without letting its presence be known is a moral blight and economic evil that threatens our institutions." *Bottom:* "An Amazing Discovery—The man in the picture is one of the vast army of rag gatherers who, from all parts of the world, are gathering raw material for 'all wool' clothing." Images and text from *The Shoddy Industry vs. the Virgin Wool Industry*, published in Chicago by the National Sheep and Wool Bureau of America in 1919. (Courtesy of Harvard Library.)

description of how to transform woolen rags into shoddy. Seven panels illustrate the process, including six shown together as a panel; all of the stages occur in a single factory setting. Different states of the process are shown: "cutting out seams from clothing, sorting out different colored rags, dusting rags, interior of duster, pressing rags into bales, interior of press, sewing up bales." Two of these images are focused very much on the idea of cleanliness and the purification of rags: the dusting process and then a close-up of the associated machine's interior.

Excluded by the frame lines and not shown are matters related to the rags' provenance—origin stories and disease pathogens alike. A woman carries them in without explanation; the effect is a kind of "blocking out." There is a presumption or implication that somehow dusting the rags is a process that cleans them. In fact, physical (as opposed perhaps to spiritual or metaphysical) purification would require either sterilization or piecemeal disinfection. The impression produced is thus a sense of cleanliness without the matter of sterilization, disinfection, or quarantine being raised. Taken as a series, the visual suggestion is that through processes of industry, there is a progression toward productive homogeneity, and this is presented as a positive and productive development rather than a dangerous and deceptive one—that the uniformity of the product renders it both harmless and commercially fecund. These images and others[61] are an effort to wipe out the other, deeply felt negative images, objects, and associations we have been exploring. The illustrations from *Scientific American* and a photograph from the same year by social documentarian Alice Austen can be interepreted as part of a single matrix (fig. 3.17).

Both these images from 1896 coexist, with neither fully adequate to the matrix of phenomena, ideas, and emotions in which they are embedded; co-constitutive of one another, they exist in tandem and are fundamental to shoddy's nature as commodity and medium. Austen's photograph is one image from a series she made over the course of several years documenting laborers and street life in New York City.

The image from *Scientific American* depicts a process that anonymizes the particularity and individuality of articles and origins, shredding them and ultimately compressing them into a homogeneous substance. The anonymizing of the rag elements is critical; they must be ground together in such a way as to elide, for future wearers and users, the previous wear-mylodon in the lowermost zone of a floor deposit of Indian refuse, and upon the recent expedition the cave earth with its "culture layer" was entirely removed for 58 feet inward from the entrance to settle beyond doubt the relation of these fossils to the Indian remains resting upon them. At this significant spot, where again the Pleistocene and recent deposits lay in contact, and where the specimens found were labeled according to their position, whether from the black (modern) earth above or the yellow (ancient) earth below, a completed examination should decide whether man had or had not encountered the tapir and mylodon in the Valley of the Tennessee.

After a visit to "Indian Cave" on the Holston River, Carrol's Cave, and the Copperas and Bone Caves, near Tullahoma and Manchester, Tennessee, a new set of conditions was presented at Big Bone Cave (one mile from left bank of Caney Fork and about two miles above its mouth in Rocky River, Van Buren County, Tennessee).* There the bones of the gigantic fossil sloth (megalonyx), still retaining their cartilages, were exhumed from a dry deposit of the refuse of porcupines and cave rats, mingled with fragments of reeds used as torches by Indians in a gallery 900 feet from the entrance, thus presenting us in the final summing up of this strange evidence a new notion of the relation of the modern Indian to this extinct

PREPARING OLD WOOLEN RAGS FOR SHODDY CLOTHING.

Shoddy consists of old woolen rags and shreds of stockings, flannels, and other soft worsted fabrics torn

CUTTING OUT SEAMS FROM CLOTHING

rate boxes, according to the color and quality of the material. The boxes are made of wood and are about 4 feet in height and about 18 inches square, and will hold about 50 pounds each. Each hand can sort about 90 pounds daily. After the stock is sorted it requires cleaning to free the material of dirt. This is performed by passing the stock through what is called a duster. This apparatus is a square boxlike structure, inside of which is a revolving wheel made of wood about 4 feet in diameter, containing four paddles, the blades of which are about 4 feet in length and about 8 inches in width. The material, to the amount of about 50 pounds, is placed in the apparatus: the paddles, which revolve at the rate of about 300 revolutions per minute, striking the rags and throwing them against the sides of the structure, which forces out the dirt, the dust being carried off at the top by means of a two-foot blower. The dusting operation takes about one minute. The stock, according to the quality and color, is then put into bins holding about 1,000 pounds each, ready for packing into bales. Where the stock is composed of old clothing or any material containing seams or patches, it is necessary to cut them out, so that the cotton can be burned out. The seams are cut out by women and girls with shears and knives, the operation for each suit taking about 10 minutes. The strips of cloth are then dusted and the cotton in the

SORTING OUT DIFFERENT COLORED RAGS.

DUSTING RAGS

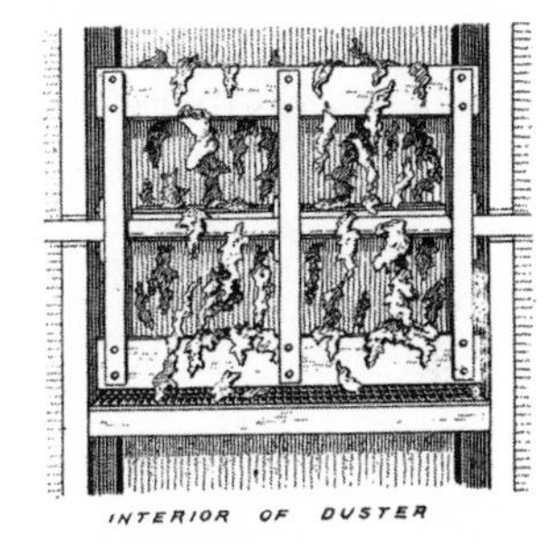
INTERIOR OF DUSTER

PRESSING RAGS INTO BALES

INTERIOR OF PRESS

SEWING UP BALES.

PREPARING OLD WOOLEN RAGS FOR SHODDY CLOTHING.

Fig. 3.16 "Preparing Old Woolen Rags for Shoddy Clothing." From *Scientific American* 75, no. 3 (July 18, 1896): 37. (Courtesy of Harvard Library.)

Fig. 3.17 Alice Austen, *An Immigrant and His Rags* (1896).
(Staten Island Historical Society/Historic Richmond Town.)

ers: their illnesses, individualities, even the environments from which they came. Then the shoddy could be stuffed into mattresses, or respun to make fresh cloth, with nary a trace of its past.[62] Through such a process, shoddy could both appear and function as productive and useful, rather than derelict and depraved, so that people could put aside questions of disease and immigration.

Shoddy's manufacture requires, in part, the active fashioning of this erasure—of making disappear the sick people, old clothes, and their wearers and handlers, the physical and emotional spaces where they came from, the people who were touching and collecting them. Its instability as a commodity results from both the constantly shifting origins of its component rags and also the inherent imperfection of such attempts at erasure; a trace will always remain. At the same time, shoddy existed as trope and fact, metaphor and material, an entity in between people and

Fig. 3.18 Alice Austen, *Lost, Strayed or Stolen* (1896).
(Staten Island Historical Society/Historic Richmond Town.)

things, disease and health, between the apparent legibility of commodities and their unknown natures. Material and emotional eradication is always imperfect, but again the trace remains.

In figure 3.18, another in the series of photographs by documentary photographer Alice Austen, also from 1896, ragpickers take a break from pushing their heavy wheelbarrows loaded with masses of compressed rags, which one imagines have been collected from all over the city. The photograph catches them lost in thought, and we, as viewers, are left to ponder commodities from malt extract to condensed soup to musical comedies that are advertised on the walls behind them. In this photograph, two men pose with old-fashioned rag carts heaped with piles of rags, great mounds of the unknown, collected from around the city to be sold, eventually, for paper and shoddy. We don't know what is within. Are they pants? Shirts? Blankets? And regardless, what is their pedigree? All that is certain for us is that they operate as currency and livelihood, in one way or another.

Although identifying marks on the bags or histories associated with them are absent, labels elsewhere in the scene are very much present. Product labeling is omnipresent in the rest of the image, taken up with a wall covered in posters. This, the background, is practically a backdrop, chock-full of commodities and of-the-moment signifiers: Pabst Malt Extract, Rochester Beer, ketchup, as well as narratives processed into modern musicals. In this context, there is something timeless about the foregrounded scene: the ragmen in vests and their old-fashioned rag carts. Even as the ragpickers collect the old rags for assimilation into the new material of shoddy, we glimpse them here subjected through the ubiquity of advertising culture to an ongoing process of cultural assimilation, the logic of which demands that they themselves eventually must also disappear.

The disconnect is startling, and provocative as well. In an irony that may or may not have been lost on the photographer (to say nothing of her human subjects), repeated one after the other, over and over, at the bottom of the wall are posters for *Lost, Strayed or Stolen*, the name of a musical, but also very much what could refer to the content of the bags, the great mounds as a whole.

Liveliness and Formlessness

Ultimately, the puzzle presented is that of the commodity itself, as living and dead, as a vital materiality of the unknown. Karl Marx had written of the fetishism of the commodity, "As soon as it emerges as a commodity, it changes into a thing which transcends sensuousness. It not only stands with its feet on the ground, but, in relation to all other commodities, it stands on its head, and evolves out of its wooden brain grotesque ideas."[63] It lives as both something tangible and physical—satisfying to a human need of whatever kind—but also as a mystery, a fetish.

Insofar as the commodity mediated objects, images, and traces of violence, could things be, as one writer put it, "shoddied to death"? "Is shoddy, in fact, anything more than a superficial symbol of a deep-seated moral disease?" What of this "villainous compound, pounded, rolled, and smoothed" that yet has "the solubility of the reflected image?"[64] And if only symbolic, soluble like an unfixed photographic print, what of its stinkiness, its utility, its proximity to the bodies of millions?

Shoddy had emerged as a vital material, exemplifying the tension between the human (body, skin, and hand) and the industrial machine, between the self and the other, and between the profane and perfected. The economy of rags illustrated the growing disillusionment with the products and by-products of industrialization, and yet also shoddy's hidden qualities of spiritual and social redemption. We return at last to "the shape of a baby," described in the epigraph to this act. Whatever the life-form buried amidst the rag bales at Perth Amboy actually was, that there was an air of putrefaction is beyond dispute, recalling Georges Bataille's formulation of the *informe*, of "that shipwreck in the nauseous," that "fetid sticky object without boundaries, which teems with life and yet is the sign of death."[65]

EPILOGUE

Shoddy Renaissance

The shoddy heap with which this book began is in some ways similar to one that might have been there a hundred or even two hundred years ago, when textile waste also made its way onto regional agricultural fields, when it had been, as described by Henry Mayhew,

> good for manure, and more especially for the manure of agriculturalists in Kent, Sussex, and Herefordshire, to the culture of a difficult plant—Hops. It is good for corn (judiciously used), so that we again have the remains of the old garment in our beer or our bread.[1]

The days of all-wool shoddy in the shoddy heaps of West Yorkshire lasted well into the last decades of the twentieth century, fertilizing not only the hops of Kent, but more proximately the abundance of rhubarb grown by E. Oldroyd and Sons and others in the Rhubarb Triangle, flanked by Wakefield, Morley, and Rothwell, and just to the east of Batley and Dewsbury. Passersby on the local network of pedestrian byways, from professional gardeners to recreational ramblers and amateur botanists, would remark upon—sometimes in awe—the unintended consequences. In those years, the shoddy heap would have two main components: scraps and leftovers from the wool-scouring industry, and a portion of the deviled old woolens ground at the shoddy mills deemed not useful for further textile manufacture. Brass army buttons would appear, battered from having passed through the teeth of the "devil" and yet having made it, almost intact, first into the heap and ultimately, through the process of the nitrogenous wool's disintegration, into the soil itself. Unusual weeds appeared, exotic species that had been brought in, in the fleeces of sheep imported largely from the former British colonies (often the same colonies as those supplying the influxes of immigrants in the second half of the twentieth century).

Both buttons and weeds became objects of fascination and nostalgia—of almost epic proportions—in the increasingly post-colonial and post-industrial landscape of West Yorkshire in the second half of the twentieth century and into the twenty-first (fig. E.1). The *Illustrated London News* ran a story in November 1951 called "The Odyssey of Some Australian Weeds Found in an English Field," describing shoddy weeds as "full of wiles" like

> Odysseus, the classic wanderer of ancient times . . . whose epithets are equally deserved by some twenty-odd kinds of weed found growing this year in a Midland English field which had travelled even further. These weeds were "escapes" from the "grey shoddy" (waste wool from the Yorkshire mills) which were being used as field manure.[2]

John Martin, whom we met in the prologue, remains fascinated with the pockmarked "shoddy buttons" that he collected during his years working in the fields. He reflects in conversation:

> In the farm where I worked, the soil was full of army buttons. I get the impression that these uniforms were just thrown completely into machines, chewed up, and reused in the woolen industry. And all of the buttons would've been part of the detritus they needed to get rid of. Every field that we used was full of 'em. And of course all army regiments have different designs on the buttons—and y'know—oh that's a different one, oh I'd not seen that one before. The buttons and the shoddy weeds have the same origins in that they're basically just a waste product from the woolens industry.[3]

Today there are many fewer weeds, and when John and I wandered the heap, he couldn't find even one to speak of. His "shoddy weed herbarium" is full of specimens, but the dates peter out in the late 1980s, the very years when the amount of wool being scoured and pulverized plummeted, and when what scoured wool was left for the fields ended up being sterilized

Fig. E.1 A collection of pockmarked buttons picked from shoddy heaps by the author, and now scattered over "where seeds from the 'grey shoddy' have grown in a railway siding." Image by Hanna Rose Shell.

first, hence driving out any life from the seeds. While the heaps remain on the edges of the agricultural fields, John is nostalgic for the days of collecting exotic "transient" species that arrived, suddenly, with a splash of vibrant color from a remote corner of what had once been an empire.

Weed collector Ida Hayward, in collaboration with a botanist, described her own fascination with the woolen waste, while calling others to do the same: "Men do not see the glory in the grey, the wonder in the common-place, the poetry in their commerce, the beauty amidst their toil, but here is a plant romance just at their door."[4] It is shoddy as poetry; shoddy as a poetic material waiting to be found in its multiple facets. Hayward's moment was in the immediate aftermath of World War I, but the dumping continued well into the late twentieth century and into the twenty-first. The "grey" has changed its constituent matter, along with its unintended and often captivating consequences; its value for contemplation, along with its poetry, remains. In the heap before us, these are piles of industrial detritus that cannot find another home; the dregs of the dregs aspiring to some form of late-in-life productive function. Most features of this heap suggest a recent vintage. While some of the material is clearly wool waste—leftovers from the scouring process carried out at one of the few remaining wool treatment facilities in the region—there is a lot of what looks like laundry lint. These striated mounds are baled-up dust collected from extractor fans in modern rag-grinding mills or collected in "cyclones," machines used to shake rags before shredding. The powdery excess of this process—modern-day "devil's dust"—is compressed into bales, shrink-wrapped, and tied with wire. Once it has been delivered to the field, the wires are clipped and the shrink-wrap cut. Some blows away across the fields, but moisture from the environment, and sometimes from a hose, helps the bales retain their form. Then there are the sequins, zippers, boa feathers, and the shredded bits of other items dotting the countryside: made-in-China T-shirts, traditional *shalwar kameez* mailed from Pakistan only to be discarded by local teens bent on assimilation, and even the occasional brass button remains of army uniforms worn in battles long ago won or lost. A modern-day facility in Batley produces dust for shoddy heaps as a by-product of mattress and carpet production. As the proprietor described it:

> The dust is sucked out of the product we manufacture from the textile waste and the dust then goes into a hopper, which holds the dust, and then daily, twice daily we mop it out, and then that dust gets bagged up and it goes to the farm. And the farm processes it through and puts it back onto the land. Probably it sits six to twelve months in a big pile in a field, mixes it with this stuff when he's got waste, vegetables or anything he's got in his waste, and it rots away and then it rots back into, and he puts it back the ground again.[5]

How much nitrogenous material is contained in that dust is hard to know. Claims as to shoddy's universally beneficial fertilizing properties don't at first glance seem to apply to its most recent incarnation. One of shoddy's first great promoters, the West Yorkshire businessman and historian Samuel Jubb, proclaimed of the "rag and shoddy system" that "there are no accumulations of mountains of debris to take up room, or disfigure the landscape; all—good, bad, and indifferent—pass on, and are beneficially appropriated."[6] Much of what we find at the heap today will never be "beneficially appropriated," for the synthetic fibers that make up an increasingly large part of the heap biodegrade much more slowly than natural ones, if they biodegrade at all. Synthetics also do not fertilize the soil the way nitrogenous wool does. It is dumped on the heap nonetheless because the dumpers have a vested interest in calling this material "useful," rather than refuse, and in depositing it as such. In doing so, they avoid the substantial waste fees levied by the United Kingdom in accordance with the very European Union legislation that drove so many in this region to support the "Brexit" referendum.

In twenty-first-century West Yorkshire, the recycling industry has largely outlived the wool trade, its technological and material progenitor. After the precipitous decline of the woolen and worsted industries in the region, slow progress has been made toward economic revitalization.[7] The region between Leeds and the nearby city of Bradford has become a locus for the collection and sorting of clothing donated to charities from all over the United Kingdom. The clothes come in to organizations with a range of social service and religious affiliations, from Ummah Welfare Trust, inspired by Islamic teaching, to the West Yorkshire Firefighters Charity, as well as from companies specializing in collecting in bulk from charity shops nationwide such as Colltex clothing recycling and exporters (fig. E.2).[8] Organizations then often pass the collection along to the for-profit companies that painstakingly sort and bale the clothes for shipment overseas, as well as to the shoddy mills that go on to supply the many local mattress and carpet pad factories in the region. Whereas for many decades this sorting and baling was carried out on-site at several specialty mills, sorting has become an increasingly rare practice in the last decade or so, a major one having burned down in 2016. Residents of Northern England who grew up working in the shoddy industry still boast of their knowledge of materials and their ability to discern—just by touch, sometimes even at a glance—wool from rayon, acrylic, or polyester.

The regional companies in the business now, however, mostly do it differently, baling the bulk of the clothing and either shipping it to Poland, where sorting can be done more cheaply, or sending it outside the EU for sale in secondhand clothing markets; and as it so happens a sizable portion of what gets sorted in Poland rather ironically ends up coming back to West Yorkshire for grinding up. As a local businessman tells me:

> The sorting started disappearing some years ago, and it's just got disappearing and disappearing, like a vanishing, where's there's nobody sorting rags these days. Three or four years ago, them clothes used to come in and they used to have twenty, thirty people with a conveyer belt, all the rags were thrown on, sorted out, different qualities, different countries where they were wanted, and left in there was the stuff no countries wanted to wear . . . winter coats that no warm countries want and that was what was for us. . . . They still get collected here—right across this parking lot, even—and then go overseas for grading: trucks, boats, trains, more trucks in one direction and then back. Then the rags return to us eventually, and we use these machines here to grind them down.[9]

The irony is not lost:

> I know it sounds silly—some days it makes me laugh, some days we are all shocked—but a lot of our product now which is coming out of here [Batley and Dewsbury remain central hubs for the collection of charity clothing donations], all that's done is it gets loaded into wagons, shipped straight out of the country; goes on to Holland, to Poland; goes because it's much cheaper labor to be sorted in other countries in the EU and then the wool and other contents is baled and brought back to us.

Whereas traditional (wool) shoddy was (and still is, to a limited extent) formed into yarn, cloth, clothing, and army blankets, this new shoddy makes its way into carpets, carpet backing, mattresses, speaker systems, and padding for automobiles and that multicolored padding in enve-

Fig. E.2 "Cash 4 Clothes." Scenes from the secondhand textile collection industry in Batley. Image by Hanna Rose Shell.

lopes that always ends up scattered every which way. Though some wool remains, as I'm reminded on multiple occasions, shoddy used in contemporary applications may have a much higher synthetic fiber content than traditional shoddy. (Some in the area choose instead to refer to it as "flock.")

Indeed, the rise of synthetics, both in how the population clothes itself and in the composition of textile waste, poses an interesting dilemma not only for shoddy's ability to fertilize the soil, but also the viability of local and global economies. Synthetic fibers, introduced in the 1920s and 1930s, would come to have a profound impact on the content of our clothes. While military garments remained wool throughout the twentieth century, slacks, jackets, sweaters and shirts, ladies' wear and menswear, and swimwear and underwear came to incorporate nylons and rayon (made from regenerated cellulose). "Artificial wool" mimicked many of the textures and qualities long ascribed only to wool, whether of the "virgin" or "reworked" variety.[10] By the late 1940s and early 1950s, industry-standard atlases of textile microphotographs refer to "old" and "new" textile fibers, with microscopy used to provide information about how to detect the presence of one or more such materials in a given swatch. "New" and "old" (wool, cotton, jute, silk, linen) are broken down further into plant, animal, mineral, and synthetically sourced. Filament and staple-rayon fibers were acetate-derived; synthetics included nylon, vinyon, Orlon, alginate; mineral fibers such as asbestos were highlighted for their ability to mimic certain qualities of wool. The new fibers would come to provide the textile substrate of articles previously wool-derived, from clothing to mattresses and insulation.[11] Instead of instructional photomicrographs to help people discern the presence of shoddy in woolen goods not labeled as such, in the new manuals, photomicrography demonstrated the corruption of worsteds by rayon and acetate, with shoddy nowhere to be found.[12]

Whereas the "virgin" wool industry had been agitating for the "proper labeling" as shoddy qua shoddy since the early twentieth century (culminating in a series of "Honest Cloth" and "Truth in Fabric" movements in the 1920s), it was only in the context of the proliferation of synthetic fibers that labeling legislation was finally passed in the late 1930s (in the United States, this was the Fabric Labeling Act of 1937 and the Wool Products Labeling of Act of 1939; in 1971 members of the European Union began to standardize labeling terminology), only after which wool-specific regulations were passed, including provisions that mandated consumer notification of the presence of shoddy (wool-specific labeling)—though under the revamped and adjectival terminology of "reworked," "reprocessed," and eventually "recycled" wool. It was the comparative ease of production that would ultimately facilitate the rise of increasingly inexpensive (and often less durable) apparel.

*

A story of globalization—of globalism and reactions to it—is written into Batley's Spa Field Mills, purpose-built in the nineteenth century for the textile industry. Inside, a company called Tom W. Beaumont Ltd. (now simply Beaumont's) once used the whole space for secondhand clothing processing and shoddy manufacture. Beaumont's—like many other textile recycling companies in the region—has been processing synthetics since the 1970s. Beaumont's used to be a one-stop shop: collecting, sorting, reselling, and also shredding discarded clothes worth more as shoddy than as shirts. Now, though, the sorting operation has been shut down, after which Beaumont's continued to use the back part of the mill to make wiping rags and polishing and cleaning cloths for the manufacturing and automotive industries. Beaumont's leases space in the front, across the parking lot, to a firm called Felt Tec, which purchased old rag-grinding lines the length of a basketball court and now manufactures insulation padding for the bedding industry.[13] Indeed, an adjacent building is leased to Harmony Beds, whose workers incorporate Felt Tec's products in their making use of Felt Tec's wares into their signature mattresses.

Meanwhile, in other parts of West Yorkshire, old woolen mills are converted to luxury apartments. As Felt Tec's co-owner Steve Carter told me: "It's going back a lot of years this shoddy, you can still see the old mills around here where it says shoddy mills on them, they're not shoddy no more, they're flats." From Carter's perspective, what's being produced now, the shredded clothes that he turns into padding, isn't "real" shoddy, which for him means detritus from wool scouring and old wool shredding. He continued:

> Now what we're doing is we've moved on a stage further. It's similar but ours is a cleaner process. If I had a handful of wool shoddy here in one hand and a hand full of ours in the other, you'd tell straightaway because it [the former] smells. It's very interesting, shoddy; it's a thing, and one I've known all my life, but it is also such an interesting word. It makes me think about so many things now; it's even a very interesting world.[14]

Henry Hardcastle, who works making audio speakers out of the same sort of contemporary version of shoddy, becomes similarly intrigued by the evocative nature of shoddy, after recounting to me his own history in the region:

> You don't know about shoddy? Well, everybody knows about shoddy. That's what it was around here, shoddy mills. Fella once fell into a machine and came out with no clothes on, just his boots. So the story goes.

He and I talk more, and he begins to think more broadly, philosophically even:

> You get me to thinking more: a lot of words derived from something else take on a completely new meaning, I don't mean, I mean if you went to a shoddy manufacturer and said his stuff was rubbish, as i.e., shoddy, he'd probably hit you, because he thinks it's absolutely marvelous what he's making, you know, so it's just the way the language has gone. That's it basically. But yeah, marvelous material made from rubbish. Wonderful stuff and hateful stuff all becoming the same thing.[15]

Amidst all this change, there are a number of senses in which shoddy has fertilized a "rebirth" beyond even whatever goes on at the heap in terms of fertilizing the crops. First of all, at least in the case of that luxury condo building, it is as something kind of hip, aiming for an eco-cool urban rejuvenation of what is one of the most poverty-stricken areas in the United Kingdom today. Meanwhile, far beyond West Yorkshire, in the high-fashion districts of London, Paris, New York, and beyond, highlighting rather than hiding the "reworked fiber" aspect of clothes is now fashionable, au courant, on trend, and marketable both at the luxury and fast-fashion levels. There is, in full swing, a "green fashion" movement and discourse of "sustainable fashion," and the hipness of it all, as well as the long history of vintage clothing and secondhand-store shopping as a fashion subculture.[16]And then there is H&M, where we began, which, despite the masses of dye pollution and landfill waste that its business model presumes, pitches itself with the marketing slogan "there's only one rule in fashion: recycle your clothes." Also of relevance today is the growing critique of the economics and cultural politics of the international secondhand clothing industry, which I had explored before embarking on this shoddy journey whose beginning and ending are here in Batley.[17]

Much deeper than fashion, and even much deeper than the reality of needing to "save the planet," is the issue of rebirth that has always been at the heart of the shoddy zeitgeist (since the early 1800s). This is the deep—both industrial and meta-spiritual—idea of "the transmutation of fibers," of "renaissance," to use the term of a nineteenth-century economist.

Now it's not wool but textiles and fibers writ large. What's really going on in Batley and Dewsbury is a way in which the transnational story of shoddy that we've been witnessing has entered a new phase, with the "first world" serving as a powerhouse for finding ways to grind up, or simply bale and sell to poorer places, all the cheap clothes made in sweatshops in China, Vietnam, and so on (by ill-treated workers who don't have the benefits of the first-world protections that were fought for so hard by the unions and also the anti-shoddy factory people in the 1910s and 1920s). And now we have that stuff in our insulation: regenerated "denim insulation," for example, or thermal emergency aid blankets.[18] While in 1920s America debates raged about the appropriateness of using reworked materials in mattresses, contemporary Batley has become a central region for the bedding industry, mattress insulation, frames, coils, and headboards. All this flock stuffing, ground-up old clothes of all fibers, laid into padding. A case in point, Pakistani and Indian immigrants are hard at work making and stacking and grinding away at the mattresses and beds at a company called Harmony Beds and another called Dreamers.[19]

Everyone and every kind of fiber is working together or perhaps against each other—grinding down in the devil of our own making, that is potentially also our salvation; things move with a vibrant materiality, embedded with the riches and the vulnerabilities that histories, identities, and corporalities instantiate. The complicated piecemeal process that is both the devil and its dust, which is also—and always has been—at the heart of the nature of the world we inhabit now and that of our future.[20] Shoddy's constantly shifting manifestation in space, time, and states of decomposition and recomposition binds nature and artifact, found and forged, wasted and wanted, in the unseen bits and inscrutabilities at the core of objects of our everyday existence. These mysteries can be a source of awe, aesthetic engagement, material archaeology, and historical analysis. In its capacity as all of these, shoddy is always already in transition, always in between states, never one thing or another, but in multiple ways and domains simultaneously both and neither. At its origins a textile woven and worn, it is simultaneously a product, a process, a material, a metaphor, and a medium of communication. As a technological artifact, shoddy is formed at the nexus of production and consumption, of waste and manufacture. An assemblage and a mangle, shoddy is always in a sense both unique (in its particular manifestation) and generic (insofar as it is a dynamic medium).[21] In it, multiple human and nonhuman voices and natures, materialities and traces, intermingle. Its existence is liminal, coming from and returning to that sometimes-uncanny space between the human body and its environmental surface layers. At levels both material and metaphorical, shoddy continues to wear the traces and bear the burdens of an environmental and political history in progress.

Shoddy as a technological process developed in parallel with other industrial as well as social and cultural "melting pot" and "mangling" technologies. And shoddy as an economic structure and cultural logic—in ways Marx anticipated and those he might not have—helped to form connections of international commerce that persist today. Yet the current political backlash by populist separatists who react, in part, against the condition of free international trade calls into question the purportedly Darwinian logic that undergirds pro-free-trade arguments. Does unfettered capitalism "evolve" the most robust society, or does it lead to diminished diversity through the introduction of "invasive species" of commercial product? And does this diminishment of diversity impoverish us culturally and imperil us environmentally? Or does shoddy and the network of cultures, economies, and technologies that it connects disclose a redemptive unifying principal? What can this scrap of shoddy, this army blanket, this recycled woolen shirt, the stuffing of this mattress—what can these tangible yet somehow materially elusive materials tell us about who we are, where we came from, and where we may go?

Acknowledgments

Substantial support for this project came from the Leo Marx Career Development Fund, the Massachusetts Foundation for the Humanities, the W. F. Milton Fund, and the Agosto Foundation.

Curators, archivists, and others at many museums, libraries, and cultural centers provided invaluable insight and assistance, in particular the American Textile History Museum (now closed), the Bangor Historical Society, Special Collections at Leeds University Library, West Yorkshire Archive Service, YIVO Institute for Jewish Research, the Batley History Group, Harvard's Baker Library and Special Collections, the Map Collection, and the Center for the History of Medicine in the Francis A. Countway Library of Medicine. I am also very grateful to the following businesses, charities, and trade groups: SMART (Secondary Materials and Recycled Textiles Association); Greenhill Textiles; Felt Tec, Ltd.; Milltex; the Garment District; Bobby from Boston; Whitehouse & Shapiro; the Ummah Welfare Trust; the Batley Central Methodist Church; Beaumont's; Harmony Beds and Mattresses; Trans-America Trading Co.; Len-Jay, Inc.; Henry Day & Sons, Ltd.; E. Oldroyd & Sons, Ltd.; Batley News; and Thomas Chadwick & Sons.

Charles Day and the late Vivien Tomlinson provided access to their personal collections and family histories in the shoddy industry. I will always hold dear the set of samples gifted to me from the cabinets in the front office of Charles's shuttered shoddy factory. The value of the perspectives of many others in West Yorkshire cannot be overstated—those of Mark Andrews, Steve Carter, Henry Hardcastle, Simon Jackson, Alistair Longbottom, John Martin, Jack Morrell, Janet Oldrohyd, Raasheed Parwashi, Brian Yelland, and especially that of local historian and Batley booster Malcolm Haigh, who first welcomed me to his town with open arms in 2011 (fig. A.1).

I thank friends, colleagues, and mentors in the Harvard Society of Fel-

Fig. A.1 Walking along Howley Mill Lane with Malcolm Haigh. Photo by Alex Wolkowicz.

lows, in the Film Study Center at Harvard, and in the School of Humanities and Social Sciences at MIT, including in the Department of History, the Department of Literature, and the Program in Science, Technology and Society; I also thank my friends and colleagues at the University of Colorado Boulder, in the Art and Art History Department, the Cinema Studies and Moving Image Arts Department, and the History Department, as wells as my MIT graduate student research assistants Renée Blackburn, Grace Kim, and Shira Shmuely.

Generosity, enthusiasm, and logistical support have abounded. In Prague, Michal Kindernay, Dana Recmanová, and Miloš Vojtěchovský; in Jihlava, Lenka Dolanová; in West Yorkshire, Andy Abbott, Jon Barraclough, Sarah Beck, Rose Borthwick, Yvonne Carmichael, Connor O'Grady, and Alex Wolcowicz; in Baltimore, the late Bill Shapiro; in Boston, Bruce Cohen, Liz Donovan, the late Bobby Garnett, Yvon Lamour, Ben Mantyla, Judy Spitzer, and Patrick Sylvain; in New York, Sarah Bruner, Jessamyn Hatcher, and the Stubins—Ed, Sunny, and Eric; in Port-au-Prince, Trenton Daniel, Gary Dauphin, and Elizabeth Pierre-Louis.

Extra special thanks to Vanessa Bertozzi for her brilliance and kindness; to Robert Buehler for their talent and commitment; to Bud Bynack,

Megan Hustad, Leo Marx, and Sina Nafjali for their ways with words; and to my editor Karen Darling and the team at the University of Chicago Press.

Earlier formulations of the project have appeared in print in the journals *History and Technology, Transition: An International Review, History and Technology*, and *Cabinet: A Quarterly of Art and Culture* and have received valuable feedback at meetings of the Society for the History of Technology, the Society for Cinema and Media Studies, and the Society for Literature, Science and the Arts, and in presentations at the Bard Graduate Center, Bar Ilan University, New York University, the University of Wisconsin–Madison, University of California, Santa Cruz, Pratt Institute, and the Parsons School of Design.

Four generations of my family have supported this project; this includes my daughter and husband, to whom this books is dedicated, as well as my parents, Marc Shell and Susan Meld Shell, my brother Jacob Shell, my aunt Andrea Meld, and my cousins in textiles the Cytrynbaums of Montreal, especially the late great Leonard Cytrynbaum (1936–2017). This book is imbued with the spirit of my late grandparents who saw Shoddy almost till the end: Sophie Cytrynbaum Shell (1923–2017), Murray Meld (1920–2017), and Sophie Kushner Meld (1916–2017). The past yet lives in the present.

Notes

PROLOGUE

1 A $2.99 reusable tote bag is available at checkout, branded with "*H&M Conscious*" and a puzzling slogan: "There's only one rule in fashion: recycle your clothes." On fast fashion and its impacts, see Cline, *Overdressed*, and Fletcher, *Sustainable Fashion and Textiles*.

2 The film *Secondhand (Pèpè)* (Third World Newsreel, 2007), was co-directed by Vanessa Bertozzi and based on a multiyear collaboration also documented in Shell's photo-essay "Textile Skin," with photographs by Vanessa Bertozzi.

3 "Use of Shoddy Is Greatest in America: Workingmen Here Literally Wearing the World's Old Clothes," *New York Times*, July 10, 1904, financial supplement, p. 4.

4 "Only Americans Wear Virgin Wool: Order Gives Our Soldiers Pure Worsted Uniforms, Though 'Shoddy' Satisfies Fighters of Other Nations," *New York Times*, February 10, 1918.

5 William Blake, "And did those feet in ancient time," from the prologue to the epic poem *Milton*. (Blake, *The Early Illuminated Books*, 205.)

6 On the "putting-out system" for wool, see Berg, *The Age of Manufactures*, 219, and on the transition to the "factory system," 225–27. Furthermore, there were simply many more people; through the course of the eighteenth century, England's population nearly doubled, leading to widespread concern over slums, the spread of disease, and rising crime rates.

7 Malcolm Haigh, interview with author, October 2012.

8 Malcolm spoke to me of ". . . immigrants and their offspring. They started coming in the 1950s. Since then they just multiplied. Now there's a little over 10,000. In fact at one time just about the whole of the taxi industry here was operated by Asians, and they're the ones who brought the shops. At one time shops used to close up at five or half past five. When they came of course they opened as long as they could." Haigh, interview.

9 Stuart, "Rags and Their Products in Relation to Health," 644–47. At the time, Stuart was Batley's chief medical officer, as well as an accomplished poet and author.

10 Towne v. Eisner, 245 US 418 (1918).

11 By the century's end, it had also become an area for farming rhubarb, both in fields and in the dark in sheds that dot the nearby landscape.

12 On this term and the emergence and contested etymologies of others used herein, and the history and significance of shoddy's emergence more generally, see Shell, "Shoddy Heap."

13 The genealogy of wool waste processing, particularly rags' reduction to fiber and respinning into new fabric for wear, is distinctive from—although technologically related to—that of the collection and reuse of cotton and linen rags for writing paper manufacture, especially in relation to issues of social and cultural meaning.

14 Parker, *Final Report of United States Liquidation Commission, War Department*, 29–30, 95.

15 In fact, by this point there had long been clear proof of what proper sterilization looked like. Still suspicion always hung in the air, especially where and whenever what was referred to in the transcript as "secondhand shoddy" was concerned.

16 Weaver v. Palmer Bros. Co., 270 US 402 (1926). Holmes spoke the dissenting opinion on behalf of Harlan Stone, Louis Brandeis, and himself.

17 Pa. Ls. 1923, c. 802 put a total ban on shoddy effective at the start of 1924, while allowing for the possibility, at the least, of the development of a valid means of sterilization for other secondhand materials: "No person shall employ or use in the making, remaking, or renovating of any mattress, pillow, bolster, feather bed, comfortable, cushion, or article of upholstered furniture: (a) any material known as 'shoddy,' or any fabric or material from which 'shoddy' is constructed; (b) any second-hand material, unless, since last used, such second-hand material has been thoroughly sterilized and disinfected by a reasonable process approved by the commissioner of labor and industry. . . . The question for decision is whether the provision purporting absolutely to forbid the use of shoddy in comfortables violates the due process clause or the equal protection clause."

18 The Supreme Court ultimately weighed the case in large part based on the applicability as precedent of the 1888 case Powell vs. Pennsylvania, 127 US

678, which had overturned an 1885 law prohibiting the sale of "oleomargarine" (Laws of Penn. of 1885, No. 25, p. 22).

19 "Bars Lifted on Shoddy (Weaver v. Palmer Bros. Co., 1925)," *Business Law Journal* 8 (1929): 128–29. Full explanation contained herein: "Pennsylvania Prohibited the Use of Shoddy in Mattresses to protect the public from unsanitary conditions in the bedding industry. Then she included in the prohibition all stuffed and filled bedding: covers, quilts, and comfortables. In 1923, when the law was enacted, a Connecticut company made 750,000 shoddy-filled comfortables and sold $180,000 worth in Pennsylvania. The manufacturer's excuse for using reclaimed wool and cotton fiber was: 'The world's supply of new wool is insufficient to clothe the people of the temperate zones and to meet other demands.'" See also Holmes, "Bars Lifted on Shoddy" (1925), in *The Dissenting Opinions of Mr. Justice Holmes*; also Frankfurter, *Mr. Justice Holmes and the Constitution*, 32–33.

20 Holmes, "The Contagiousness of Puerperal Fever."

21 From "The Rag-Picker's Wine," in Baudelaire's *Fleurs du Mal* (1857): "One sees a rag-picker go by, shaking his head, / Stumbling, bumping against the walls like a poet, / And, with no thought of the stool-pigeons, his subjects, / He pours out his whole heart in grandiose projects." In the original: "On voit un chiffonnier qui vient, hochant la tête, / Butant, et se cognant aux murs comme un poète, / Et, sans prendre souci des mouchards, ses sujets, / Epanche tout son coeur en glorieux projets."

22 Since its establishment 1936, the IABFLO (International Association of Bedding and Furniture Law Officials) has been responsible for coordinating state-level enforcement of bedding label laws and specifications, enacted, it is argued, to give consumers notice about what they are unlikely ever to see directly, as well as if a mattress has been previously used.

23 "On these premises, if the Legislature regarded the danger as very great and inspection and tagging as inadequate remedies, it seems to me that in order to prevent the spread of disease it constitutionally could forbid any use of shoddy for bedding and upholstery." (Weaver v. Palmer Bros. Co., 270 US 402, 415 [1926].)

24 Weaver v. Palmer Bros. Co., 270 US 402, 403–4 (1926).

ACT ONE

1 Carlyle, "Signs of the Times." On clothing as ever-shifting in relation to social relations, see Braudel, *Capitalism and Material Life, 1400–1800*, 226.

2 Jubb, *The History of the Shoddy-Trade*. On the earlier history of secondhand clothing trade in England, see Lemire, "Shifting Currency."

3 Some have said it derives from an Arabic word meaning to reuse, whereas others posit origins in the word *shod* (an archaic past tense of *shed*). Alistair Longbottom, interview with author, Greenhill Mills, Batley, November 1, 2012; and Malcolm Haigh, interview with author, Batley, September 1, 2011; see also Jubb, *The History of the Shoddy-Trade*, 31; Megraw, *Textiles and the Origin of Their Names*; and "The Woollen and Worsted Trade of Great Britain."

4 Jenkins and Ponting, *The British Wool Textile Industry, 1770–1914*, 92–102.

5 Leo Marx has written eloquently about Thomas Carlyle and situated him at the start of the development of a line of critical thinking about technology, as the developer of the term "age of machinery" and also the term "industrialism." See Marx, *The Machine in the Garden*; "Technology"; and "The Idea of 'Technology' and Postmodern Pessimism."

6 Marx, *The Machine in the Garden*, 170.

7 Carlyle, "Signs of the Times."

8 Berg, *The Age of Manufactures*, 40–44; "The Woollen and Worsted Trade of Great Britain."

9 Landes, *The Unbound Prometheus*, 41–45.

10 Berg, *The Age of Manufactures*, 102–5. On the woolen and worsted industries' modern emergence and dominance in the period, see Jenkins, "The Western Wool Textile Industry in the Nineteenth Century."

11 Heaton, *The Yorkshire Woollen and Worsted Industries*; Morrell, "Wissenschaft in Worstedopolis."

12 Before the mid-nineteenth century, cotton and linen rag paper was generally produced sheet by sheet, using rag pulp created in small water-powered mills. This preceded what is generally considered to be the "modern" or "industrial" era of paper manufacture marked by a transformation from rag- to wood pulp–based manufacture for paper in the mid-nineteenth century, with rags increasingly reserved for banknotes and specialty papers; Reynard, "Unreliable Mills"; Magee, *Productivity and Performance in the Paper Industry*;

McGaw, *Most Wonderful Machine*, 96–103; Andrews, *Rags*.

13 Dana, *A Muck Manual for Farmers*, 142–43.

14 Lemire, "Peddling Fashion"; Jones and Stallybrass, *Renaissance Clothing and the Materials of Memory*. These practices predated the medieval period and the Renaissance, as described by Wild, "The Romans in the West," 93.

15 On making do, see Strasser, *Waste and Want*, 72–73. The concept of "bricolage," introduced by Claude Lévi-Strauss (in his 1962 *La pensée sauvage* [The Savage Mind]), is employed in the service of attempts to refocus attention in the history of technology on practices of use, repair, and recycling. See, for example, Männistö-Funk, "The Crossroads of Technology and Tradition," 756. On the turn toward use, repair, and innovation more generally in the history of technology, see Edgerton, *The Shock of the Old*, and Zimring's monographs *Cash for Your Trash: Scrap Recycling in America* and *Aluminum Upcycled: Sustainable Design in Historical Perspective*.

16 Linebaugh, *The London Hanged*, 247–55. Also see Fontaine, *Alternative Exchanges*.

17 On the idea of waste in nineteenth-century England, see Cooper, "Modernity and the Politics of Waste in Britain," and Scanlan, *On Garbage*. As a concept operating in capitalism, as taken up by Friedrich Engels as well as by Karl Marx, see Foster, *Marx's Ecology*, 138–39; also Herman E. Daly reflects on "devil's dust" as it plays a unique role in Marx's theorization of the commodity and the economic system: "Many of the material inputs in metabolism are economic products, and some outputs of metabolism are economic products, and some outputs of metabolism are generally, not totally degraded and thus can be further consumed—for example, manure fertilizer and carbon dioxide. But the ultimate physical output of the economic process *is* totally degraded matter-energy, in Marx's term, 'devil's dust.'" (Daly, "On Economics as a Life Science," 396.)

18 Hudson, *The Genesis of Industrial Capital*; also Griffin, *A Short History of the British Industrial Revolution*, 86–104.

19 Burrows, *A History of the Rag Trade*, 1; Radcliffe, *Woollen and Worsted Yarn Manufacture*, 63–65.

20 McCulloch, *A Dictionary, Practical, Theoretical, and Historical, of Commerce and Commercial Navigation*, 1437.

21 On the development of shoddy, and for evidence of the mythology surrounding its development, see, for example, "The Woollen and Worsted Trade of

Great Britain," 57; "Yorkshire," *Westminster Review* 71, no. 140 (April 1859): 179–95; also Jenkins and Ponting, *The British Wool Textile Industry, 1770–1914*, 3.

22 An alternate account, partially detailed later in this section of the text, attributes the original invention of shoddy to a London-based Jewish secondhand clothes dealer (named Davis) during the Peninsular War—a time of high demand for military goods customarily made from Spanish wool (then under embargo). The dealer came up with the idea of producing fabric made from mixing new wool with old blankets and white flannels torn up by curry combs. When the market for such fabric later fell, he offered it to upholstery shops and saddleries, where Law himself first encountered it. See Lock, *Spons' Encyclopaedia of the Industrial Arts, Manufactures, and Raw Commercial Products*, 2058.

23 The details of the shoddy origin story of Benjamin Law were relayed to me in multiple interviews of residents of West Yorkshire, but with most clarity by local historian Malcolm Haigh. Haigh, interview, May 31, 2011, and September 1, 2011. It is also recounted in Haigh's *The History of Batley*.

24 Old wool had for centuries been put onto fields at the year's end; during the winter, the action of wind, rain, and snow would cause nitrogen to leach the wool into the soil. Dana, *A Muck Manual for Farmers*; also Winter, *Secure from Rash Assault*, 40–61.

25 Clapp, *An Environmental History of Britain since the Industrial Revolution*, 201–2.

26 Steve Carter, interview with author, October 2012.

27 On issues of repair and retrofitting of machinery as it pertains to technological innovation, see Edgerton, "From Innovation to Use" as well as his *Shock of the Old*. Recent monographs placing such focus on specific industries include Jørgensen, *Making a Green Machine*, and Zimring, *Cash for Your Trash*.

28 Burrows, *A History of the Rag Trade*; also, Alistair Longbottom, interview with author, October 2012.

29 In this sense, the shoddy industry was the creator of new kinds of jobs. Here it might be argued that the industry differed from the more established textile industries, whose mechanization triggered the widely publicized Luddite machine-thrashing protests in the early 1810s. In 1860, of the 600 people in Batley estimated to be employed by the dozens of shoddy firms in the region, 80 worked as rag grinders. But a far greater number (an estimated

500 people) were "pickers"—female rag sorters, overseen by foremen. For the causes of these protests and an astute analysis of the implication thereof, see Bailey, *The Luddite Rebellion*; employment numbers from Jubb, *The History of the Shoddy-Trade*. On the larger context of Luddism, including in Batley and Dewsbury, refer to Thompson, *The Making of the English Working Class*, 474–75, 546–50.

30 "A Shoddy Business," *The Archive Hour*, BBC Radio 4, November 13, 2004.

31 Polanyi, *The Tacit Dimension*, 21. Volume 5 of Diderot and D'Alembert's *Encyclopédie, ou dictionnaire raisonné* vividly illustrates these practices of sorting in the article on "Paper Manufacturing."

32 Ferrar Fenton, letter to the editor, *Batley Reporter*, December 1, 1880. Fenton was a Batley-based shoddy cloth and manure businessman, inventor, and translator. His letter was written in response to Edwin Law, "The Discovery and Early History of the Shoddy and Mungo Trades," *Batley Reporter*, November 13, 1880. Text transcribed from microfilm by Wendy Rose and reproduced by Maggie Land Blanck, August 9, 2019, http://www.maggieblanck.com/Land/WR.html.

33 Fenton, letter to the editor, *Batley Reporter*, December 1, 1880.

34 Head, *A Home Tour through the Manufacturing Districts of England*, 144–45, 147.

35 Head, 145. He continues, in what seems an obvious reference to Carlyle, whose *Sartor Resartus* had appeared in serial form in 1833–34: "Those who delight in matters of speculation have here an ample field provided they feel inclined to extend their researches on this doctrine of the transmigration of coats; for their imagination would have room to roam in unfettered flight, even from the blazing galaxy of a regal drawing room down to the cellars and lowest haunts of London, Germany, Poland, Portugal, &c. as well as probably to other countries visited by the plague."

36 Jubb, *The History of the Shoddy-Trade*, 20.

37 The majority of the rags processed in the shoddy towns of West Yorkshire in the first half of the nineteenth century came from Germany and Holland and—closer to home—Scotland, Ireland, and elsewhere in England. Specific numbers can be found in Bischoff, *A Comprehensive History of the Woollen and Worsted Manufactures*, 179–80, and further fleshed out in "Woollen Rags: An account of the number of bags of woollen rags imported from January 1818 to January 1822" (April 24, 1822) and "An account of the quantity

of woollen rags imported yearly since 1828; and, so far as can be stated, its proportional application to manufacture and agriculture" (*Parliamentary Papers*, 52). See also Clapp, *An Environmental History of Britain since the Industrial Revolution*, 195.

38 See Ginsburg, "Rags to Riches"; Lemire, "Consumerism in Preindustrial and Early Industrial England"; and Lemire, "Shifting Currency." Also see Shell, "A Global History of Secondhand Clothing." On the development of markets for a range of secondhand consumer goods in Western Europe, see the collection edited by Stobart and van Damme, *Modernity and the Second-Hand Trade*. For the British context, see Lambert, "Cast-off Wearing Apparell."

39 Mayhew, *London Labour and the London Poor*, 369. See also Scanlan, "In Deadly Time," and Herbert, *Culture and Anomie*, 204–52.

40 Jubb, *The History of the Shoddy-Trade*.

41 "The Batley Rag and Shoddy Sales," 37.

42 "Yorkshire," *Westminster Review* 71, no. 140 (April 1859): 191.

43 Head's description of shoddy was, for example, republished in anthologized form in Philadelphia in 1836, as well as making its way into subsequent accounts published as late as the 1880s and 1890s.

44 Head, *A Home Tour through the Manufacturing Districts of England*, 147.

45 Chambers, "Devil's Dust," 103.

46 The focus of such sentiments on rag-grinding machinery in particular is connected to the embodied and deeply personal nature of clothing, over and above any organic connotations of raw wool. Shell, "Textile Skin."

47 For an especially resonant literary manifestation, see Carlyle, *Sartor Resartus*, esp. chapter 5 of book 1 and chapter 6 of book 3. Leo Marx productively situates Carlyle's work more generally in the context of an emerging philosophy of the machine in *Machine in the Garden*, 161–72, 286.

48 Harley, "Trade," 176–90.

49 *Parliamentary Debates*, House of Commons, 24 February 1842, vol. 60, cols. 1018–82, 1065 (Fifth Day/Corn Laws, Mr. Villier's Motion, Total Repeal, Adjourned Debate).

50 See, for example, the entry for "devil's dust" in Hotten's *Slang Dictionary* of 1869.

51 See Morley, *The Life of Richard Cobden*, vol. 1, 224.

52 The transcript continues: "They now put scarcely any wool into their yarn, only just as much as will keep the devil's dust together. The rags, as you

know, are collected from the most filthy holes in London and Dublin, and are brought from the most unhealthy regions, infected by the plague and every epidemic, and of course, they are full of deadly poison."

53 *Parliamentary Debates,* House of Commons, 60, cols. 1018–82, 1066. On Ferrand's explosive use of rhetoric during his tenure in the House of Commons, see Henderson, "Industrial Legislation," 275–76.

54 "Devil's Dust," *The Spectator*, March 12, 1842, 16; Simmonds, *Waste Products and Undeveloped Substances* (1873), 26.

55 Cf. "down with one's dust," a colloquial expression meaning "to lay down one's money." See *Cassell's Dictionary of Slang* (1998; 2005).

56 From William Fox's lecture delivered at Covent Garden Theater on January 25, 1844, "The Corn Laws and Compromise." See Hirst, *Free Trade and Other Fundamental Doctrines of the Manchester School,* 180.

57 "Dissenters, Religious Duty and Religious Practices," *Common Sense, or Every-Body's Magazine* 2, no. 5 (May 1843): 114–15.

58 Bradford, *Disraeli,* 116–17.

59 Engels's focus on evaluating the state of society through the clothing worn by the population was surely influenced by his own experience as a textile industrialist and heir himself. First published in German in 1845 and translated into English in 1885 for authorized English-language publication in 1887, *The Condition of the Working-Class in England in 1844* was finally published in London in 1891.

60 Engels, 66. The passage continues, ". . . But among very large numbers, especially among the Irish, the prevailing clothing consists of perfect rags often beyond all mending, or so patched that the original colour can no longer be detected. Yet the English and Anglo-Irish go on patching, and have carried this art to a remarkable pitch, putting wool or bagging on fustian, or the reverse—it's all the same to them. . . . Ordinarily the rags of the shirt protrude through the rents in the coat or trousers. They wear, as Thomas Carlyle says, 'a suit of tatters, the getting on or off which is said to be a difficult operation, transacted only in festivals and the high tides of the calendar'" (66–67).

61 Engels, *The Condition of the Working-Class in England in 1844,* 66–67.

62 Karl Marx, "A Meeting," *Neue Oder-Zeitung,* no. 141 (March 20, 1855), 98; first published in English in the collection *Articles on Britain* (Moscow: Progress Publishers, 1971), 229–32. Reproduced in Marx and Engels, *Collected Works,* vol. 14, *1855–1856,* 98–101.

63 Marx, "A Meeting," 100.

64 Jubb, *The History of the Shoddy-Trade*, 22.

65 Jubb, 23–24.

66 Cooper, "Peter Lund Simmonds and the Political Ecology of Waste Utilization in Victorian Britain."

67 Simmonds, *Waste Products and Undeveloped Substances; or, Hints for Enterprise in Neglected Fields* (1862). The second edition, completely rewritten, appeared eleven years later under the title *Waste Products and Undeveloped Substances: A Synopsis of Progress Made in Their Economic Utilization during the Last Quarter of a Century at Home and Abroad* (1873). Simmonds was so inspired by shoddy, among other waste products, that he lobbied for and successfully helped to establish stand-alone exhibitions on the subject.

68 Jubb, *The History of the Shoddy-Trade*, 24.

69 Head, *A Home Tour through the Manufacturing Districts of England*, 141–52.

70 Black and Black, *Black's Picturesque Guide to Yorkshire*, 346.

71 This created a situation, exemplified by the discourse surrounding the "Shoddy Temple," in which the working class brushed elbows with the mercantile elite. Given this new mode of potential interaction, a man that heretofore would have had little hope of advancement might now find himself impressing his employers and other higher-ups as worthy of consideration for promotion. Haigh, interview, September 1, 2011.

72 Chambers, "Devil's Dust," 104. For the larger literary context in which shoddy became romanticized in this period, see Stuart, *The Literary Shrines of Yorkshire*, 30, 74.

73 Pickering, *The Mangle of Practice*, 6. The quotation continues: "My suggestion is that we should see science (and, of course, technology) as a continuation and extension of this business of coping."

74 Carlyle came from rural southwest Scotland and was raised with a strict religious upbringing; he had planned a career in the church, but while at university in Edinburgh, he discovered German idealism. Greatly influenced by German thinkers, especially Goethe, he began writing literary and philosophical reviews, and became an early voice of the Victorian era. Always moving between creative and philosophical worlds, Carlyle was also widely seen as a revolutionary force in his own time. Key scholarship on Carlyle includes Kaplan, *Thomas Carlyle*, and Tennyson, *Sartor Called Resartus*.

75 "*Die Kleider, ihr Werden und Wirken* (Clothes, their Origin and Influence)*: von Diog. Teufelsdrockh, J.U.D. etc. Stillschweigen und Co*[gnie.] *Weissnichtwo*, 1831." (Carlyle, *Sartor Resartus* [2000 ed.], 6.)

76 See editor's note, *Sartor Resartus* (2000 ed.), 245. See also Stowell, "Teufelsdröckh as Devil's Dust," 31–33. In changing the name (which had appeared in early letters as "Teufelsdreck"), Carlyle may have been trying to "subtilize" the term, transforming it from a general name, referring to the work as a whole, into a "proper" name, referring to a character. See Tennyson, *Sartor Called Resartus*.

77 Carlyle, *Sartor Resartus*, appendix 1, 227.

78 *Sartor Resartus* was initially published anonymously, in monthly installments in the London-based journal *Fraser's*. It appeared between November 1833 and August 1834, with gaps in January and May. A private issue of all of these sections together appeared in 1834, and then it was published in the United States.

79 In arguing for its publication, Carlyle notes that he has previously referred to the work as a "Satirical Extravaganza on Things in General." But one that well reflects his thoughts: "It contains more of my opinions on Art, Politics, Religion, Heaven Earth and Air, than all the things I have yet written." See Appendix 1 with his note to Fraser.

80 Carlyle writes of clothes as being emblems, and of emblems as all being clothes. See book 1, chapter 11, "Prospective," 56–57: "sham Metaphors . . . overhanging that same Thought's-Body . . . its false stuffings, superfluous show-coats (*Putz-Mäntel*), and tawdry woollen rags." The emblematic is perhaps the visible aspect, but is the sham the material?

81 *Sartor Resartus*, book 1, chap. 1, "Preliminary," 4.

82 "Have we not a Doctrine of Rent . . . ? Man's whole life and environment have been laid open and elucidated; scarcely a fragment or fibre of his Soul, Body, and Possessions, but has been probed, dissected, distilled, desiccated, and scientifically decomposed . . . every cellular, vascular, muscular Tissue glories in its Lawrences, Majendies, Bichats [scientific discoverers]." (*Sartor Resartus*, 4.)

83 This aspect of Carlyle's work is pointed out by Wilson, *Adorned in Dreams*, 55, and Hollander, *Seeing through Clothes*, 450.

84 Quoted by Marx, *The Machine in the Garden*, 86.

85 I here set aside Carlyle's treatment of Judaism and the Jews, including his richly ambiguous depiction of Monmouth Street (the famous neighborhood of the Jewish rag-gathering trade), later echoed by Charles Dickens in "Meditations in Monmouth-Street" included in *Sketches by Boz* (1836).

86 Disraeli, *Sybil*, book II, chap. 10, 86–87.

87 Disraeli, 87.

88 Disraeli claimed to be descended on the paternal side from Sephardic Jews of noble descent.

89 *Capital*, vol. 1, chap. 8, "Constant Capital and Variable Capital."

90 Peter Stallybrass approaches meanings inherent in the wearing of clothing for Marx in relation to his theories of value through an exploration of the notion of wrinkles as "memories." Stallybrass, "Marx's Coat," 196.

91 *Capital*, vol. 3, part 1, "The Conversion of Surplus-Value into Profit and of the Rate of Surplus-Value into the Rate of Profit," chap. 5, "Economy in the Employment of Constant Capital." Foster makes a very convincing argument as to the deeply ecological framing of Marx's argument in chapter 5, section 4, in Foster, *Marx's Ecology*, 169.

92 On the productive loop between production and consumption, and the value of waste as part of this loop in Marx's overall thought, see Gabrys, "Shipping and Receiving," 292–93.

93 Disraeli described his position in an April 3, 1872, lecture in Manchester; see Goddard, "'A Mine of Wealth'?," 281.

94 *Capital*, vol. 3, part 1, chap. 5, sec. 4.

95 See sec. 4: "This entire line of economies arising from the concentration of means of production and their use on a large scale has for its fundamental basis the accumulation and co-operation of working people, the social combination of labor. Hence it has its source quite as much in the social nature of labor as the surplus-value considered individually has its source in the surplus-labor of the individual laborer."

96 Consider, for example, the ecological economist Herman Daly, who describes "the ultimate physical output of the economy" as "totally degraded matter-energy, in Marx's term 'devil's dust.'" As Geller has noted, "Lumpe" was the nickname of the central Jewish figure in the influential anti-Semitic German play that was produced just prior to the Hep-Hep riots of 1819. See Geller, *The Other Jewish Question*, 192–97.

97 Marx, *The Eighteenth Brumaire of Louis Bonaparte*, 41.

98 Marx and Engels, *The German Ideology* (1845–46), vol. 1, part 3, sec. 6, no. A.

99 On the role of "devil's dust" in later struggles among socialists, see the remarks of Michael Bakunin, Marx's and Engels's influential anarchist opponent: "To me the flower of the proletariat is not, as it is to the Marxists, the upper layer, the aristocracy of labor, those who are the most cultured, who earn more and live more comfortably than all the other workers. Precisely this semi-bourgeois layer of workers would, if the Marxists had their way, constitute their *fourth governing class*. This could indeed happen if the great mass of the proletariat does not guard against it. By virtue of its relative well-being and semi-bourgeois position, this upper layer of workers is unfortunately only too deeply saturated with all the political and social prejudices and all the narrow aspirations and pretensions of the bourgeoisie. Of all the proletariat, this upper layer is the least social and the most individualist. By the *flower of the proletariat*, I mean above all that great mass, those millions of the uncultivated, the disinherited, the miserable, the illiterates, whom Messrs. Engels and Marx would subject to their paternal rule by a *strong* government—naturally for the people's own salvation! All governments are supposedly established only to look after the welfare of the masses! By flower of the proletariat, I mean precisely that eternal 'meat' (on which governments thrive), that great *rabble of the people* (underdogs, 'dregs of society') ordinarily designated by Marx and Engels in the picturesque and contemptuous phrase Lumpenproletariat. I have in mind the 'riff-raff,' that 'rabble' almost unpolluted by bourgeois civilization, which carries in its inner being and in its aspirations, in all the necessities and miseries of its collective life, all the seeds of the socialism of the future, and which alone is powerful enough today to inaugurate and bring to triumph the Social Revolution." See Bakunin, "The International and Karl Marx."

100 Marx, *Capital*, vol. 1, 592–93. In a footnote at this point, Marx references a report on the rag trade from 1866: Bristowe, "The Report on the Rag Trade and Its Influence in Spreading Disease," 196–208. See Act III for more on Bristowe, and matters of disease and contagion.

ACT TWO

1 Gernsheim, *A Concise History of Photography*, 55–56.

2 Bruce, *The Twentieth Regiment of Massachusetts Volunteer Infantry, 1861–1865*, 9–15.

3 Marx, "Technology: The Emergence of a Hazardous Concept," 573.

4 Oliver Wendell Holmes's father had himself given an evocative description of the rather skin-like nature of the photographic material onto which a negative was transferred into a positive image: "A sheet of the vest linen paper is dipped in salt water and suffered to dry. Then a solution of nitrate of silver is poured over it and it is dried in a dark place. This paper is now sensitive; it has a conscience, and is afraid of daylight. Press it against the glass negative and lay them in the sun, the glass uppermost leaving them so for from three to ten minutes. The paper, having the picture formed on it, is then washed with the solution of hyposulphite of soda, rinsed in pure water, soaked again in a solution of hyposulphite of soda, to which, however, the chloride of gold has been added, and again rinsed. It is then sized or varnished. . . . For, ho! when the sensitive paper is laid in the sun under the negative glass, every dark spot on the glass arrests a sunbeam, and so the spot of the paper lying beneath remains unchanged; but every light space of the negative lets the sunlight through, and the sensitive paper beneath confesses its weakness, and betrays it by growing dark just in proportion to the glare that strikes upon it." Holmes, "The Age of Photography."

5 Risch, "Supply Difficulties in 1861," in *Quartermaster Support of the Army*, 339.

6 Heaton, *Yorkshire Cloth-Traders in the United States, 1770–1840*.

7 Haigh, *The History of Batley*, 11–15.

8 Shaw, "Slave Cloth and Clothing Slaves."

9 Beckert's *Empire of Cotton* provides a rich history of cotton. Sections on the Civil War lay out vividly the extent to which cloth, and the presence or absence of the fibers required for its production, was at the core of interlinked political and economic factors that both triggered and profoundly affected the war. "Negro cloth" is touched on briefly in Beckert and Rothman's introductory chapter to their edited volume *Slavery's Capitalism*, 2.

10 By the 1830s and 1840s, such mills would have been getting some of the wool rags from rag collectors, who had long been supplying the papermaking

industry. The paper industry could only use cotton and linen, and now the wool-based leftovers had a market of their own.

11 Nevins, *The War for the Union*, 255.

12 Peace Dale Manufacturing Company Records, Baker Library Special Collections, Harvard Business School. Also Peace Dale Mill of the J. P. and R. G. Hazard Co. in Rhode Island—either they sent the cloth south for the pregnant or elderly slaves to sew into rough uniforms, or it was done cheaply by old women up north and then shipped down. See Zakim, *Ready-Made Democracy*, 51–52.

13 Problems with contracting and faulty uniforms were not entirely new with the Civil War; indeed, ten years earlier, during the Mexican-American War, there had been a host of unexpected deficiencies resulting from decisions about government contracting. (Risch, "Logistical Problems of the Mexican War," in *Quartermaster Support of the Army*, 255.)

14 Isaac Comstock, to the Van Wyck Committee Commission, February 3, 1861, Committee Papers of the Select Committee on Government Contracts, Records of the U.S. House of Representatives, 1789–2011, Record Group 233, National Archives Building, D.C., http://research.archives.gov/description/4667678. A more typical ratio would be 1:1.

15 "Notes of the Rebellion: Shabby Uniforms," *New York Times*, July 20, 1861.

16 Peace Dale Manufacturing Company Records, Baker Library Special Collections, Harvard Business School.

17 Beckert's *Empire of Cotton* provides an exceptionally rich account of this. For a concise version, see Cole, *The American Wool Manufacture*, vol. 2, 496.

18 So derisively nicknamed in the South, which was forced by means of exorbitant import tariffs to purchase goods manufactured in the North rather than in Britain, as they had previously. See Taussig, *The Tariff History of the United States*, 70–89.

19 Described in "Specifications on Patents Recently Filed," *Mechanics Magazine* 66 (January 3, 1857): 16.

20 "Patent Rag Shredding Machines," 225–26.

21 Shaw, *Homefront and Battlefield*, 22.

22 Cole, *The American Wool Manufacture*, vol. 1, 377; on the industry's rapid mechanization, see Cole, vol. 2, 494–95.

23 A 1902 report issued by Congress cited figures showing that between 1860 and 1870, capital expenditure in the shoddy sector increased by 561 percent

and that the value of sector output increased 339 percent. ("*Shoddy vs. Pure Wool*: Hearings before the Committee on Ways and Means, House of Representatives, Fifty-Seventh Congress, First Session," 11.) On the increasing role of shoddy in woolen manufacture during the Civil War, see Cole, *The American Wool Manufacture*, vol. 1, 315–16, 377; and on its percentage of the total output of wool manufacture after 1860, see Cole, vol. 2, 205.

24 Bigelow, *An Address upon the Wool Industry of the United States*, 10.

25 Brandes, *Warhogs*; on contracting and problems in this early period in the war, see 67–75. Also see Wilson, *The Business of Civil War*, esp. 24–25. Wilson describes uniform scandals in the two most populous states, Pennsylvania and New York, as early as the beginning of July, citing a satirical July 6 *Vanity Fair* cover image, with the heading "Pennsylvania Volunteers," that showed soldiers in ragged uniforms that fall to pieces on their first wearing.

26 The committee's charge was as follows: "Resolved, That a committee of five members be appointed by the Speaker to ascertain and report what contracts have been made by any of the departments for provisions, supplies, and transportation; for materials, and services, or for any articles furnished for the use of government without advertising for proposals, as required by the statute of 1861; the parties to whom given; the compensation and terms thereof, and the reasons therefor. Also, where proposals were received, if contracts were awarded to the lowest bidder; if not, the reason therefor. Also, whether the contracts are in accordance with the specifications inviting proposals; and if any alterations, the reasons for the same. Also, whether any person or persons have any interest in the contracts thus made and awarded, or obtained the same, or profits therefrom, except the contractors. That said committee shall have power to send for persons and papers to administer oaths and examine witnesses, and report at any time." Described in Bolles, *The Financial History of the United States, from 1861 to 1885*, 230–31, with the specific quotation from the 37th Congress, Session 1, reporting from July 8, 1861, p. 45.

27 Wilson, *The Business of Civil War*, 152.

28 "What Next?," *New York Times*, July 26, 1861.

29 "As the war got underway, enormous expenditures for supplies required by the military inevitably led to reports of abuses of the procurement system and to calls in Congress for an investigation into the situation" (National Archives Record 22.43 37A-E21.1). For details on the Van Wyck Committee's

activities, see Weigley, *Quartermaster General of the Union Army*, 195–200. Testimony on the subject of flaws and corruption in the system of government contracting was a vast undertaking and now fills more than 2,000 pages of a congressional report. (37:2 House Rep. 2 [serial 1142, 1143].)

30 Interview in Philadelphia, March 6, 1862, included in vol. 1143, p. 880.

31 Benjamin, "The Work of Art in the Age of Mechanical Reproduction," in *Illuminations*, 226.

32 These terms are taken from the trichotomy proposed in the semiotic system of Charles Sanders Peirce, as laid out in his "What Is a Sign? Three Divisions of Logic." (Peirce, *Philosophical Writings of Peirce*.) The Peircean theory of signs (index, icon, symbol) has since been applied in the realm of photographic analysis and the philosophy of photographic practices, for example, Krauss's "Notes on the Index: Part 1."

33 Tomes, "The Fortunes of War," 228. Roland Barthes captured and recast the philosophical points of André Bazin, Walter Benjamin, and even Peirce, in his eloquent *Camera Lucida*. Consider, too, in this regard, the comments of Holmes's own father's famous prewar essay for the *Atlantic Monthly*: "The Stereoscope and the Stereograph" (1859); on the elder Holmes's perspective on war photographs in particular, see Holmes, "Doings of the Sunbeam," 11.

34 37th Congress, Session 1, reporting from July 12, 1861, p. 71.

35 "Mr. Shoddy[,] having made much Money through Contracts, is invited to an Evening Party."

36 Morford, *The Days of Shoddy*.

37 Olcott, "Song of the Shoddy," 142.

38 "Shoddy: The Way It Is Made," *Scientific American* 5, no. 15 (October 12, 1861): 228; cf. "Shoddy: Its Material and Manufacture," *Scientific American* 15, no. 13 (September 22, 1866): 202.

39 The article continues its value as a source of jobs for many who otherwise would not be employable in the textile industry ("Shoddy: The Way It Is Made").

40 "Shoddy: Its Material and Manufacture," 22.

41 Reported on and summarized as "The Blanket Question" (*Vanity Fair*, November 16, 1861). This article makes jest at other commentary coming into newspapers about the blankets; letters of self-defense had been sent on various occasions by those in charge of filling blanket contracts. For example, "Letter of October 31, 1861 from Davies to Brig.-Gen L. Thomas," reproduced

in the *New York Times*, November 2, 1861.

42 From a letter dated July 24, 1861, to Brigadier General Francis Laurens Vinton from Quartermaster Montgomery C. Meigs (Meigs Papers, Library of Congress), quoted at greater length by Nevins, *The War for the Union*, 291.

43 Weigley, *Quartermaster General of the Union Army*, 253.

44 "The Blanket Question," *Vanity Fair*, November 16, 1861.

45 The stripping bare of bodies is described particularly vividly here: "Corpse-robbing, a repellent, but extremely common habit, led to a large proportion of the Confederate army wearing Federal light blue trousers and captured boots; so popular did this practice become, that soon Confederate regiments were wearing Union head-dress and jackets as well." (Haythornthwaite, *Uniforms of the Civil War, 1861–65*, 103.)

46 A number of scholars have noted a connection between anger against war profiteering and anti-Semitism. (Jews were historically associated with the rag and clothing industry, and frequently connected in the popular mind with sharp and unethical business practices.) On this topic, see Bunker and Appel, "'Shoddy,' Anti-Semitism and the Civil War"; Glanz, *The Jew in Old American Folklore*; and Handlin and Handlin, "A Century of Jewish Immigration to the United States." More recently, see Mendelsohn, *The Rag Race*, 159–82.

47 Twain and Warner, *The Gilded Age*.

48 On the new currency of such terms, see Alexander de Tocqueville's diary from 1861, specifically the entry from Saturday, August 17, p. 254. Publications from the period describe "a nouveau riche and a parvenu." For example, "A Great Social Problem," *Continental Monthly* 6, no. 5 (October 1864): 443; for a very early use of "nouveau riche," see *Daguerreotype* 3 (1849): 80.

49 See Disraeli, *Sybil*, in which a character named "Devilsdust" plays a pivotal role (see act 1).

50 Near the war's end, the *New York Herald* reported on the so-called "shoddy people" with a much milder dose of derision, labeling them as basically assimilated with the elite, or alternately running head to head with them. "Our élite, our aristocracy of money, our shoddy people, have run their made race of extravagance and show at the fashionable watering places, and are returning to commerce in the city of unparalleled display." Comment reprinted in *Prairie Farmer* 14, no. 17 (October 22, 1864): 267, shortly after the devastating losses on the Union side in the Battle of Cedar Creek; cited in Nevins, *The War for the Union*, vol. 3, 216.

51 Reed, *The Russian Ball*, 11, 30–31.

52 Sanitary fairs were organized throughout the Northern states by chapters of the United States Sanitary Commission, founded in 1861 as an umbrella organization for benevolent societies focused on the collection of funds and materials (clothing, shoes, and bandages especially) so as to positively impact the health and well-being of soldiers on the front. Though the precise causes of illness, and infection in particular, remained unknown to most, the association of dirt and rot with illness and disease was clear to many if not most observers on the front, and would have come across vividly in letters written home to family members. For context, see Brown, *Foul Bodies*, 366; also Frederickson, *The Inner Civil War*.

53 Nevins, *The War for the Union*, vol. 3, 317–19.

54 Shaw, *Homefront and Battlefield*, 77.

55 Druitt, *The Principles and Practice of Modern Surgery*, 498.

56 Livermore, *My Story of the War*, 121–26. This engagement with the "blanket question" during the Civil War, with blankets being thought of as possessing a kind of material and moral agency beyond themselves, is redolent of the narrative—often expounded and often disputed—about New England settlers of the 1750s and '60s who purposefully traded blankets with native communities specifically to expose them to smallpox. On dirty blankets as a weapon of war, see Fenn, *Pox Americana*, 80–88; and Brown, *Foul Bodies*, 129–31.

57 See the discussion in the *American Journal of the Medical Sciences* 45 (1863): 268. A very similar reference appears in the *Boston Medical and Surgical Journal* 67 (1862–63): 198.

58 Walt Whitman's "The Wound-Dresser" (1865) was later set to music by the American composer John Adams. Whitman's personally documented experience as a hospital attendant was a visually searing one, as suggested by his note on first entering an army hospital just after the Battle of Bull Run: "Outdoors, at the foot of a tree, I notice a heap of amputated feet, legs, arms, hands, etc., a full load for a one-horse cart. Several dead bodies lie near, each covered with its own brown blanket. In the dooryard are fresh graves, mostly of army officers, their names on pieces of barrel staves or broken boards, stuck in the dirt." (Quoted in Meltzer, *Walt Whitman*, 98.)

59 See, for example, Grant, *Oliver Wendell Holmes Jr.*; also see White, *Justice Oliver Wendell Holmes*, 72.

60 This is the case even though it has since been shown to how great an extent

many such images were staged, produced through what Trachtenberg refers to as "stagecraft" in "Albums of War," 7.

61 Notable examples here would include Davis, "'A Terrible Distinctness'"; Lee and Young, *On Alexander Gardner's Photographic Sketch Book of the Civil War*; also Trachtenberg, "Albums of War."

62 Frassanito, *Gettysburg*; also see Frassanito, *Early Photography at Gettysburg*.

63 Orvell, *The Real Thing*, 96.

64 See Harris, "'Truthful as the Record of Heaven.'"

65 Generally, Gardner and his team took a series of exposures, both stereographic and single-camera, at the scene. Gibson probably handled the stereographic camera, whereas O'Sullivan likely took the single-view series.

66 I see the "textile skin" concept as contributing to the range of approaches to considering themes of iconicity and epistemology of photography, including Mnookin, "The Image of Truth"; Tucker, *Nature Exposed*, 53–55; and Leja, *Looking Askance*, 21–58. For photographs as historical "fragments," see Didi-Huberman, *Images in Spite of All*, 80–81.

67 Frassanito, *Gettysburg*; also see Frassanito, *Early Photography at Gettysburg*.

68 The relevant section of *Gardner's Photographic Sketch Book* opens with "Gettysburg, the scene of Lee's defeat in 1863, is a post borough and . . . stands on elevated ground, in the midst of a fertile farming country. The Court-House and public offices are handsome and commodious; and the private residences are all built in a neat and substantial manner. The town has a flourishing Lutheran Theological Seminary, with a library of about ten thousand volumes; and is also the seat of Pennsylvania College. The manufacture of carriages is carried on to a greater extent than any other business. A number of copper mines have been opened in the neighborhood, and worked with considerable success. The town numbers about four thousand inhabitants."

69 Barthes, *Camera Lucida*, 90–91.

70 For details on blankets, their production and supply, and related issues during the war, see Risch, *Quartermaster Support of the Army*, 357–59.

71 My allusion here is to Bazin's seminal "The Ontology of the Photographic Image," wherein he describes: "Those grey or sepia shadows, phantom-like and almost undecipherable, are no longer traditional family portraits but rather the disturbing presence of lives halted at a set moment in their duration, freed from their destiny; not however by prestige of art but by the power of an impassive mechanical process: for photography does not create

eternity, it embalms time, rescuing it simply from its proper corruption."

72 By way of example, consider this description of a particular blanket sold by antiquarian dealer Dave Taylor's Civil War Antiques. Close reading indicates the extent to which its valuation is its connection to a distinct and discrete soldier's body—in this case a Levi S. Vandycke of the 142nd Regiment. "A Real Civil War US Army Issue Blanket ID'd to a New York Soldier: Standard army pattern utilizing very loose, coarsely woven, twill (diagonal weave) wool with plenty of 'shoddy' mixed in. This being bits and pieces of scrap wool used by makers to cut costs. About 80 by 70 inches (they were roughly 80″ × 65″ fresh off the looms). We have gained a few inches of stretch after 150 years. Standard unbound ends with dark stripes running across the width. And the magic part . . . the loosely woven 'US' in the center executed in brown wool yarn. Brownish gray overall with darker brown end stripes 2¼ inches wide running across the blanket 4–6 inches from each end, closer on the bottom because of edge wear and fraying. Large U.S. in brown yarn about the same color as the end stripes, and measuring 4 inches by 4 inches. Wear spots about the size of the palm of one's hand to right of the US, and a couple of others on the lower left, some areas of thinning fabric and several small holes, but generally very solid, able to be handled and displayed open or rolled. . . . This blanket surfaced decades ago in upstate New York and has a family note on a scrap of brown paper probably written in the mid-20th century, identifying it as 'Uncle Lee Van Dyke's Civil War Blanket.' The name checks out as Levi S. Vandycke, who served in the 142nd New York from September 1862, to September 1864, and was wounded 7/6/64 at Petersburg. The regiment saw some hard fighting, losing 3 officers and 126 men killed or mortally wounded. They served in several different corps, including the 11th, 10th, and 18th, and saw action at Drewry's Bluff, Bermuda Hundred, Cold Harbor, and Petersburg during Vandycke's period of service. They saw their last fighting at Fort Fisher in 1865, but that was after Vandycke had been discharged, perhaps as a result of his wounds. A rare issue item. $2,850.00 SOLD." The decrepitude of the weave, combined with the note "yellowed and with holes, somewhat tattered," only serve to reinforce its authenticity and thus its value for collectors. (Text as described at Dave Taylor's Civil War Antiques, http://www.angelfire.com/oh3/civilwarantiques/1411webcat.html.)

73 Interview with Matt Bishop, curator of the Bangor Historical Society, June 2016.

74 Sontag, *On Photography*, 15.

75 Bazin, "The Ontology of the Photographic Image," 14. On an expansion of this analysis of the Shroud of Turin that puts things more directly in dialogue with semiotic notions of presence and iconicity, see Didi-Huberman's "Index of the Absent Wound (Monograph on a Stain)."

76 Oliver Wendell Holmes Sr. republished "The Stereoscope and the Stereograph," first printed in the *Atlantic Monthly* in 1859, as part of his essay collection *Soundings from the Atlantic* (124–65), in which his other essays on photography (including "Doings of the Sunbeam" and "Sun-Painting and Sun-Sculpture") were also included.

77 An interesting contrast might be made between this figuration of the American flag as more-or-less ragged or noble, and the political slogan from the post–Civil War US of "waving the bloody shirt," a phrase used by the North dismissively as a way to quell and put down Southerners who seemed to use the death of their soldiers and the pain caused by the war as a means of justifying various Reconstruction-era political causes. See Budiansky, *The Bloody Shirt*, 1–4. Budiansky describes not only the false tale of the collection and waving of Huggis's bloody shirt on the floor of Congress in 1871, but also retellings of the beating of Sherman in terms of the fashioning of the bloody shirt as a "relic."

78 Orvell, *The Real Thing*.

79 The first US bunting-making mill was established in March 1865, by Benjamin Butler, Union army general, politician, and industrialist, as well as notable abolitionist, whose factories in Lowell, Massachusetts, had supplied blankets and other supplies to soldiers throughout the war. It was only at its end, however, that he went into bunting; prior to that he had been a military blanket maker. (Hearn, *When the Devil Came Down to Dixie*.)

80 General Benjamin F. Butler incorporated the United States Bunting Company in March 1895 and started producing bunting material for use in American flags. Butler already had substantial ties and connections in Congress and was able to use this to his advantage to put lucrative government contracts in place, well in advance of when he had sold a single flag. In 1866 he donated an oversize flag of his own making to the US Senate to fly over the Capitol. See Hayes, *The Fleece and the Loom*, 62.

ACT THREE

1 From the redirect examination of Joseph H. Raymond. (William Lockwood and Emory McClintock v. Bartlett et al., section 222, p. 74.)
2 Withington, "Transmission of Infectious Diseases through the Medium of Rags," 37–38.
3 It is from this position that it ultimately served as a medium for scientific work and experimentation, as in the well-documented case of the studies made on yellow fever in Panama in the early twentieth century. See Sutter, "Tropical Conquest and the Rise of the Environmental Management State," 317–26, 605–6.
4 "Miasma" refers to the Greek for "pollution." The word's use in medical literature became widespread after its usage in *De noxiis paludum effluviis*, published in 1717, authored by Italian physician Giovanni Maria Lancisi. For further context, see Parker, *Miasma*, and "Miasma Theory of Disease," in Zimring and Rathje, *Encyclopedia of Consumption and Waste*, vol. 1, 539–41.
5 This idea of "essential seeds" left open the idea that these might be living seeds, or if not living themselves, then somehow a kind of proto-life-form that could be responsible for contagious illness such as the various outbreaks and manifestations of plague over the preceding and into future centuries. An 1882 medical dictionary defined fomites as "substances capable of retaining contagium-particles and thus or being the means of propagating any infectious disease." (Quain, *A Dictionary of Medicine*.)
6 For an early instance of Fracastorius (Latin for Fracastoro) being credited as the inventor of what would come to be called "germ theory," see Garrison, "Fracastorius, Athanasius Kircher, and the Germ Theory of Disease"; also "Germ Theory of Disease," in Zimring and Rathje, *Encyclopedia of Consumption and Waste*, vol. 1, 317–19.
7 McNeill, *Plagues and Peoples*.
8 Nohl, *The Black Death*, 94–95.
9 *Loimologia, or, an historical Account of the Plague in London in 1665, With precautionary Directions against the like Contagion* was published in 1720 as an English edition of Hodges's original *Loimologia, sive, Pestis nuperæ apud populum Londinensem grassantis narratio historica*, from 1672, which was translated by apothecary and medical writer John Quincy, who also added an additional essay about means to combat disease in France at that time.

10 See Hodges, *Loimologia*, second section, *Of the Cause of a Pestilence, and a Contagion*, 30: "AND for what concerns that Pestilence now under Enquiry, this we have as to its Origin, from the most irrefragable Authority, that it first came into this Island by Contagion, and was imported to us from *Holland*, in Packs of Merchandice; and if any one pleases to trace it further, he may be satisfied by common Fame, it came thither from *Turkey* in Bails of Cotton or Silk, which is a strange Preserver of the pestilential Steams. For that Part of the World is seldom free from such Infections, altho' it is sometimes more severe than others, according to the Disposition of Seasons and Temperature of Air in those Regions: But if any would yet more intimately be acquainted with its Origin, it concerns him to know all the Changes the Air in these Climates is subject to, and its various Properties of Dryness, Moisture, Heat, Cold, *&c.*"

11 Hodges, 262–63, 273–74.

12 Aspects of Dickens's, Engels's, and Mayhew's understanding of disease causation, in particular as it pertains to issues of smell and decomposition, are covered in Johnson, *The Ghost Map*, 127–31.

13 The famous episode occurred during the extended cholera epidemic that ravaged London at midcentury. Snow traced the 1854 outbreak at Broad Street, "The Cholera Near Golden Square," as he called it, to an "index" case in which a diarrhea-soaked cloth diaper had been deposited into the system directly adjacent to the water pump. Snow, "Cholera and the Water Supply," 12. See Snow, *On the Mode of Communication of Cholera*, 2nd ed.; also Vinten-Johansen et al., *Cholera, Chloroform, and the Science of Medicine*. "In the four to five day interval between her child's onset of diarrhea on August 28–29, 1854 and subsequent death on September 2, 1854, Mrs. Lewis had soaked the diarrhea-soiled diapers in pails of water. Thereafter she emptied the pails in the cesspool opening in front of her house. Likely baby Lewis had *Vibrio cholerae* which contaminated the napkin used to absorb diarrhea."

14 Snow, *On the Mode of Communication of Cholera.*

15 Reach, *Fabrics, Filth and Fairy Tents*, 17.

16 Head, *A Home Tour through the Manufacturing Districts of England*, 145.

17 Innis, *The Skin, in Health and Disease*, 44.

18 Engels, *The Condition of the Working-Class in England in 1844*, 72.

19 Snowden, *Naples in the Time of Cholera*, 15–16.

20 Isaac Claesz. van Swanenburg's *The Removal from the Wool of the Skins and*

the Combing (1595) is one of a series of seven paintings he made in the late sixteenth century, four of which are close studies of specific stages in the processing of wool.

21 "'Racks' are, or I may say were, used in drying and stretching the woollen goods formerly manufactured in the neighbourhood. The operative part of the woollen trade carried on within this city was entirely confined to the finishing [of] the pieces for use, after they had been spun and wove. From the warehouses within the city the raw materials were distributed into the neighbouring villages, and then returned in the piece. Here the pieces were submitted to a variety of processes, as washing, milling, fulling, dyeing, raising, cutting, hot-pressing, and packing; they were rack-dried after the four first operations." (Shapter, *The History of the Cholera in Exeter*, 183.)

22 The illustrations in Shapter's volume, three included here, are the work of Exeter painter and lithographer John Gendall.

23 Closer to the sixteenth century, when Fracastoro was writing, objects also were burned during periods of disease. During various plagues, although cities would be shut down to prevent the seepage of sickly air outside and quarantined, within those quarantined areas, articles of clothing were collected and either burned or buried.

24 The International Sanitary Conference was first held in 1851, after which thirteen more took place, the last in 1938. The first and second were held in Paris, in 1851 and 1859.

25 Rosenberg, *The Cholera Years*; Kraut, *Silent Travelers*, 33.

26 It is notable that the report would in fact be cited by Marx and from it Marx quoted extensively in his own discussion of rags in volume 3 of *Capital*.

27 Hayes, *The Fleece and the Loom*.

28 Simon and the Great Britain Privy Council, *Public Health: Eighth Report of the Medical Officer of the Privy Council, with Appendix, 1865*. This text was modified only slightly and incorporated into volume 6 of the *Reports of the United States Commissioners to the Paris Universal Exposition of 1867*, 135.

29 National Association of Wool Manufacturers, "Statistics of the British Wool Manufacture," 57.

30 Bristowe, "The Report on the Rag Trade and Its Influence in Spreading Disease," 204.

31 Best practices for sterilization were widely disputed at the time, with steam heat, chemical, and water-based methods all in play and practicality and

efficacy often at odds; as Bristowe concluded: "Even if it had been shown that infectious diseases were frequently disseminated by means of rags, it would have been difficult to suggest a remedy which would be at the same time efficient and practicable." Barring disinfection prior to a previous owner's handing off used clothes to a retail dealer, and rendering it a felony to pass rags along with previous connections to the sickly, he foresaw "no practical means of dealing with the matter; the use of disinfectants, or of any process for purifying rags, either at the rag merchants or in the [shoddy or paper] mills, would be attended (so far as I can ascertain), not only with great inconvenience, but with considerable expense; and any such addition of expense, in the present condition of the rag . . . trades, would be in a high degree injurious to these branches of industry" (Bristowe, 205).

32 Baldwin, *Contagion and the State in Europe, 1830–1930*, 164–86.

33 Joseph Lister describes the application and efficaciousness of carbolic acid in this regard in his 1870 article in *The Lancet*, "On the Effects of the Antiseptic System of Treatment upon the Salubrity of a Surgical Hospital, Part 1," 4–6.

34 Simon and the Great Britain Privy Council, *Public Health: Eighth Report of the Medical Officer of the Privy Council, with Appendix, 1865*, 19.

35 Most of the International Sanitary Conferences focused on cholera. Creating international standards for dealing with people and goods coming from cholera-infested places was front and center, even as anthrax and yellow fever grew to be ever-larger concerns; and in the case of rags, smallpox was often the most proximate concern.

36 On the state of regulation of the importation of old clothing into the United States in the early 1870s, see "The Rag Trade," *New York Times*, July 30, 1871.

37 A typical example of concerns about mattresses in general was shared at Sanitary Conferences in regard to an epidemic of cholera that flared in Naples in 1884. Once the epidemic was under way, public health officials burned or disinfected mattresses and clothes and held up ships that had any non-disinfected items on board. Sailors fleeing Naples amidst the crisis tossed goods overboard in haste; they threw corpses and mattresses into the bay. A fisherman salvaged a mattress and brought it home, and all of his family members were infected. This mattress, which was itself made of rags, was discarded as part of a kind of disinfection process; and yet, once salvaged and recovered, it ultimately became the source of a whole new epidemic in a different town.

38 "Shoddy Bedding," *Annals of Hygiene* 11, no. 2 (February 1896): 125–26.

39 This also coincided with the establishment of the American Public Health Association in 1872, and the founding of that organization's journal, *The Sanitarian*. The issue of concerns about disease in relation to the importation of rags is included in "The Rag Trade," *New York Times*, July 30, 1871. By the 1890s, it was felt by many that disinfection and quarantine requirements were causing an undue burden, as, for example, in the 1896 Surgeon General's Report, 451–57.

40 Rosenberg, *The Cholera Years*; Arnold, "Cholera and Colonialism in British India." For an exceptionally florid description of the perceived origin of the illness, and the broader context of mid-nineteenth-century attempts at understanding its transmission, see Macpherson, *Epidemic Cholera*.

41 George Miller Sternberg ultimately would argue that rags and textile waste should be considered separately from other goods in terms of sanitation. (Sternberg, *Disinfection and Individual Prophylaxis against Infectious Diseases*, 27.) A year prior, William M. Smith, health officer for the Port of New York, described in the *Sanitarian* a report, thereafter to be included in the city record and provided to the New York City Board of Health, "Contagious Diseases Propagated by Rags, and the Necessity of Disinfection." Among other things, the report discussed the comparative values of superheated steam and sulfurous acid as disinfectants, also arguing that all foreign rags should be disinfected. The point was made that so-called "Ragpickers' Disease as well as Cholera and Small-Pox could indeed be Communicated by Clothing." (*The Sanitarian*, December 1, 1885.)

42 Comparable to this is the current practice in the United Kingdom of sending away "waste" through the payment of a "waste tax."

43 From the American edition of Quain, *A Dictionary of Medicine*, 705.

44 "Filthy Imported Rags," *New York Times*, August 2, 1884; "Barring Out the Rags," *New York Times*, August 9, 1884; "The Disinfection of Rags," *New York Times*, December 18, 1884. See also Withington, "Transmission of Infectious Diseases through the Medium of Rags," part of the 1886 report of the *Massachusetts State Board of Health*.

45 "Killing Disease Germs: Another Forward Step in Rag Disinfection," *New York Times*, October 17, 1885. Sternberg, considered a founder of American bacteriology and an expert on disinfectants, described the situation with rags in detail.

46 Boil the bale, and then how are you going to be sure that every part of the bale gets boiled and heated? And even if it does, what are you going to do with this giant wet bale? If you are using chemical treatment, how are you going to get the chemicals all the way into the bale?

47 It was resolving this issue that finally convinced Britain to sign on to an International Sanitary Conference in 1892.

48 Sykes, *Public Health Problems*, 177.

49 “Cholera and Rags,” *British Medical Journal* 2, no. 1705 (September 2, 1893): 553.

50 Given how long it takes for rags to get by sea from their point of origin to England or to the United States, one can hardly be sure who wore what clothing under what conditions and where they did so.

51 “Cholera and Rags.” Passage reprinted in *Literary Digest* 7, no. 21 (September 2, 1893): 10.

52 Enacted August 3, 1882, as “An Act to Regulate Immigration,” 22 Stat. 214.

53 26 Stat. 1084 added an exclusion on persons suspected of carrying contagious diseases or those associated with such persons, or of “carrying” such things.

54 This image first came to my attention through Markel, *Quarantine!*, 89.

55 The stock character of the Jewish rag dealer is on evidence throughout the early modern period, though it is largely in the nineteenth century that it became widespread and transmitted in the form of images, poems, and various other expressions of derision.

56 Concerns about disinfection and sterilization were somewhat different in regard to making paper out of cotton and linen rags versus making shoddy clothing out of woolen rags. Through the 1860s and into the 1870s, almost all paper was made from ground-up cotton and linen rags (see act 1). Wood-pulp paper started being made in the 1860s and 1870s, and although this meant lower-end publishing would move in that direction, the production of rag paper, and hence the collecting and sorting of requisite fiber-based rags, continued. In regard to issues of disinfection, however, the situation with the treatment of cotton and linen rags was much clearer cut than with wool rags and shoddy production; the very process of pulping the rags was always also a process of disinfection. What this meant was simply by virtue of being ground-up and pulped, the commodity would be sterile (though health problems could certainly still afflict rag collectors and rags sorters) by the time the papermaking process was complete. Carbonizing (the process that

"mixed" animal and vegetable fibers, i.e., "union cloth") was an inherently sterilizing process. (See the 1906 photograph [fig. 3.10] of the "carbonizing baking oven.") But again one would never know whether particular wool shoddy had gone through such a process or not. (Sometimes manufacturers would even conceal such information, since carbonized fibers would further decrease the tensile strength of fibers.)

57 According to an eminent scholar of American ethnicity, "More than any other social or political theory, the rhetoric of Zangwill's play shaped American discourse on immigration and ethnicity, including most notably the language of self-declared opponents of the melting-pot concept." (Sollors, *Beyond Ethnicity*, 66).

58 In fact, Ralph Waldo Emerson had used the expression "smelting pot" in his journal in the 1840s, but it was with Zangwill's play that the term became "melting" and entered public discourse.

59 Such ideas of smoke-screening, on the one hand, and of moral and cultural contagion, on the other, became a driving impulse behind the rise of the "virgin wool" movement, as well as the campaign for "honest cloth" and "truth in fabric," and were also at the very origin of the labeling and branding of "virgin" and "adulterated" textile goods.

60 Shoddy interests also repositioned the industry in terms of its own moral regeneration. See, for example, *The Truth in Reworked Wool* (New York: Atlas Publishing, 1920), which described in its foreword as "having been conceived in the spirit of 'Sartor Resartus.' The manufacturer of reworked wool has been stripped to its nakedness and rehabilitated in a rational and scientific manner."

61 For example, see Kittredge, "Shoddy: Or the History of a Woolen Rag" from 1906, in which a similarly industrial and germ-free vision of shoddy is presented and heavily promoted. Throughout shoddy is presented as an industrial marvel.

62 "Preparing Old Woolen Rags for Shoddy Clothing," *Scientific American* 75, no. 3 (July 18, 1896): 37.

63 Marx, *Capital*, vol. 1, 163.

64 Tomes, "The Fortunes of War," 227–28. E. P. Whipple echoed such articulations on shoddy's meaningful materiality in "Shoddy," *Atlantic* 27, no. 161 (March 1871): 338–48, also included in Whipple's *Success and Its Conditions* (Boston: Osgood, 1871).

65 Bataille, *The Accursed Share*, 81. Bataille, *An Essay on General Economy*, vol. 2, *History of Eroticism* (New York: Zone Books, 1991). Bataille's understanding of such a "fetid object," in all its glory (a sort of "toxic melting pot," one might say), is connected to his idea of "*l'informe*" (the formless), itself a vital component of a theory of "base materialism." On form, *l'informe*, and formlessness in the context of base materialism, see Bois, "The Use Value of 'Formless'" and "Base Materialism," in Bois and Krauss, *Formless*.

EPILOGUE

1 Mayhew, "Street-Sellers of Second-Articles," in *London Labour and the London Poor*, 152.
2 "The Odyssey of Some Australian Weeds Found in an English Field," *Illustrated London News*, November 3, 1951.
3 John Martin, interview with author, October 2012.
4 Hayward and Druce, *The Adventive Flora of Tweedside*, xxvi.
5 Steve Carter, interview with author, October 2012.
6 Jubb, *The History of the Shoddy-Trade*, 24.
7 On shifts in the local shoddy industry in the twentieth century, as human-made fibers spread into the textile industry, see the minute books of the Huddersfield and District Mungo and Shoddy Manufacturers Association (1918–1968) renamed the Fibre Reclaimers' Federation in 1968 (records 1968–1978, refs. 20D81/48–49 within Records of the Wool Textile Manufacturers Federation at the West Yorkshire Archive Service Bradford); also see *Dewsbury: The Official Handbook*. On the acceleration of the decline leading up to the 1970s, see Moor, "Batley at Work."
8 The donations manager at the Ummah Welfare Trust collection point gave me some context: "What clothes actually sell in the store is donated by the local community; during Ramadan the extent is very large. In terms of clothes, we tend to take everything, anything and everything, all kinds of clothes and we organize it and sort it. And how we sort it is, if we can sell the clothes in the store, they're good enough, we do sell in the store and that helps the charity, otherwise we put them in the clothing bank and we actually recycle them—they get torn up to shreds, lots of companies around here buy them but I don't know what for—and the money that we get from recycling the

clothes that goes toward our administration costs of the charity." (Rasheed Parwashi, interview with author, October 2012.)

9 Carter, interview.

10 Rayon was developed from the first artificial silk, which had been initially invented in the 1850s, but was relatively unsuccessful on account of its flammability as well as extreme expense. "Artificial Wool" was patented in 1940. (April 23, 1940, patent #2,197,896 concerns "a new synthetic wool" produced in such as a way as to "crimp, curl or crinkle various straight natural and artificial fibers in order to obtain a product similar to wool.")

11 Von Bergen and Krauss, *Textile Fiber Atlas*; Mauersberger, *Matthews' Textile Fibers.*

12 Rainnie, "The Woollen and Worsted Industry"; Hague, "The Man-Made Fibers Industry."

13 A formerly competing outfit in Batley, called Greenhill Mills (shown in figure P.1), continued sorting in-house and supporting a vibrant community of employees until it closed due to a fire in 2016.

14 Carter, interview.

15 Henry Hardcastle, interview with author, October 2012.

16 For example, the designers profiled in Caroline Baumann's exhibition at the Cooper Hewitt Museum: Brown, Baumann, and McQuaid, *Scraps: Fashion, Textiles, and Creative Reuse*; also Fletcher and Grose, *Fashion and Sustainability*; for a larger context, see Gregson and Crewe, *Second-hand Cultures.*

17 Haggblade, "The Flip Side of Fashion"; Hansen, "Helping or Hindering?"; Brooks, "Stretching Global Production Networks"; compare with Farrant, Olsen, and Wangel, "Environmental Benefits from Reusing Clothes."

18 See Norris, "Economies of Moral Fibre?," which also discusses the Indian shoddy industry.

19 The Heavy Woollen District has one of the largest Muslim populations in the UK, including a very strong Gujarati Muslim community, which has grown up in response to a series of labor shortages that began in the 1950s. (Price, "Immigrants and Apprentices.") Estimates today put the Muslim population of Batley and Dewsbury at about 60 percent Indian and 40 percent Pakistani. The largest purpose-built mosque in the country, housing its largest seminary, the Markazi Masjid, is just down the block from the shoddy mill of Henry Day and Co., which gave me the shoddy samples (loose fiber and finished cloth, an army khaki), depicted in plate 3.

20 While there is something to be said for the metaphor of the fully multicultural interwoven tapestry, there is also something to be said for the important but vanishing trade of the sorter, through the skills of whom the integrity of different fabrics were maintained. Embedded in the value of sorting seems a kind of metaphor for separatist ideology, whereas blending has within it a spirit of collectivism or universalism.

21 Pickering, *The Mangle of Practice*, 23.

Works Cited

Andrews, Arthur E. *Rags: Being an Explanation of Why They Are Used in Making Paper*. Mitteneague, MA: Strathmore Paper Co., 1928.

Arnold, David. "Cholera and Colonialism in British India." *Past & Present*, no. 113 (1986): 118–51.

Bailey, Brian J. *The Luddite Rebellion*. Stroud, UK: Sutton, 1998.

Bakunin, Mikhail, "The International and Karl Marx." In *Bakunin on Anarchism*. Edited by Sam Dolgoff, 286–320. New York: Knopf, 1980.

Baldwin, Peter. *Contagion and the State in Europe, 1830–1930*. Cambridge: Cambridge University Press, 2005.

"Barring Out the Rags." *New York Times*, August 9, 1884.

"Bars Lifted on Shoddy (Weaver v. Palmer Bros. Co., 1925)." *Business Law Journal* 8 (1929): 128–29.

Barthes, Roland. *Camera Lucida: Reflection on Photography*. Translated by Richard Howard. London: Fontana Paperbacks, 1984.

Bataille, Georges. *The Accursed Share*. Vols. 2 and 3. Trans. Robert Hurley. New York: Zone Books, 1992.

Bataille, Georges. *An Essay on General Economy*. Vol. 2, *History of Eroticism*. New York: Zone Books, 1991.

"The Batley Rag and Shoddy Sales." In Samuel Jubb, *The History of the Shoddy-Trade: Its Rise, Progress, and Present Position*, 35–37. London: Houlston and Wright, 1860.

Baudelaire, Charles. *Les Fleurs du Mal*. Paris: Auguste Poulet-Malassis, 1857.

Bazin, André. "The Ontology of the Photographic Image." In *What Is Cinema?*, Vol. 1, translated by Hugh Gray, 9–15. Berkeley: University of California Press, 1967.

Beckert, Sven. *Empire of Cotton: A Global History*. New York: Knopf, 2014.

Beckert, Sven, and Seth Rockman, eds. *Slavery's Capitalism: A New History of American Economic Development*. Philadelphia: University of Pennsylvania Press, 2016.

Benjamin, Walter. *Illuminations: Essays and Reflections*. Edited by Hannah Arendt. New York: Schocken Press, 1986.

Berg, Maxine. *The Age of Manufactures: Industry, Innovation, and Work in Britain, 1700–1820*. Totowa, NJ: Barnes & Noble, 1985.

Bigelow, Erastus B. *An Address upon the Wool Industry of the United States: Delivered at the Exhibition of the American Institute in the City of New York, October 5, 1869*. New York: S.W. Green, 1869.

Bischoff, James. *A Comprehensive History of the Woollen and Worsted Manufactures and the Natural and Commercial History of Sheep, from the Earliest Records to the Present Period*. London: Smith, Elder and Co., 1842.

Bischoff, James. *The Wool Question Considered: Being an Examination of the Report from the Select Committee of the House of Lords*. Leeds: Edward Baine & Son, 1828.

Black, Adam, and Charles Black. *Black's Picturesque Guide to Yorkshire*. Edinburgh: Adam and Charles Black, 1862.

Blake, William. *The Early Illuminated Books*. Edited by Morris Eaves, Robert N. Essick, and Joseph Viscomi. Princeton, NJ: William Blake Trust/Princeton University Press, 1998.

"The Blanket Question." *Vanity Fair* 4, no. 99, November 16, 1861.

Bois, Yves-Alain, and Rosalind E. Krauss. *Formless: A User's Guide*. New York: Zone Books, 1997.

Bolles, Albert Sidney. *The Financial History of the United States, from 1861 to 1885*. New York: D. Appleton, 1886.

Bradford, Sarah. *Disraeli*. New York: Stein and Day, 1983.

Brandes, Stuart D. *Warhogs: A History of War Profits in America*. Lexington: University Press of Kentucky, 1997.

Braudel, Fernand. *Capitalism and Material Life, 1400–1800*. New York: Harper Torchbooks, 1975.

Bristowe, John Syer. "The Report on the Rag Trade and Its Influence in Spreading Disease." *Eighth Annual Report of the Medical Officer of the Privy Council with Appendix, 1865*, 196–208. 1866.

Brooks, Andrew. "Stretching Global Production Networks: The International Secondhand Clothing Trade." *Geoforum* 44 (2013): 10–22.

Brown, Kathleen. *Foul Bodies: Cleanliness in Early America*. New Haven, CT: Yale University Press, 2009.

Brown, Susan, Caroline Baumann, and Matilda McQuaid. *Scraps: Fashion, Tex-*

tiles, and Creative Reuse. New York: Cooper Hewitt, Smithsonian Design Museum, 2016.

Bruce, George Anson. *The Twentieth Regiment of Massachusetts Volunteer Infantry, 1861–1865*. Boston: Houghton, Mifflin, 1906.

Budiansky, Stephen. *The Bloody Shirt: Terror after Appomattox*. New York: Viking, 2008.

Bunker, Gary L., and John Appel. "'Shoddy,' Anti-Semitism and the Civil War." *American Jewish History* 82, no. 1 (January 1, 1994): 43–71.

Burrows, Hermann. *A History of the Rag Trade*. London: Maclaren, 1956.

Carlyle, Thomas. *Sartor Resartus*. Boston: James Munroe and Company, 1836.

Carlyle, Thomas. *Sartor Resartus: The Life and Opinions of Herr Teufelsdröckh in Three Books*. Edited by Mark Engel and Rodger L. Tarr. Berkeley: University of California Press, 2000.

Carlyle, Thomas. "Signs of the Times." *Edinburgh Review* 49 (June 1829).

Chamberland, Charles E. *Le charbon et la vaccination charbonneuse d'après les travaux récents de M. Pasteur*. Paris: B. Tignol, 1883.

Chambers, Robert. "Devil's Dust." *Chambers's Journal of Popular Literature, Science, and Art*, February 16, 1861.

"Cholera and Rags." *British Medical Journal* 2, no. 1391 (August 27, 1887): 478.

"Cholera and Rags." *British Medical Journal* 2, no. 1705 (September 2, 1893): 553.

Clapp, B. W. *An Environmental History of Britain since the Industrial Revolution*. London: Longman, 1994.

Cline, Elizabeth L. *Overdressed: The Shockingly High Cost of Cheap Fashion*. New York: Portfolio/Penguin, 2013.

Cole, Arthur Harrison. *The American Wool Manufacture*. Cambridge, MA: Harvard University Press, 1926.

Cooper, Tim. "Modernity and the Politics of Waste in Britain." In *Nature's End: History and the Environment*, edited by S. Sörlin and P. Warde, 247–72. New York: Palgrave Macmillan, 2009.

Cooper, Timothy. "Peter Lund Simmonds and the Political Ecology of Waste Utilization in Victorian Britain." *Technology and Culture* 52, no. 1 (January 2011): 21–44.

Daly, Herman E. "On Economics as a Life Science." *Journal of Political Economy* 76, no. 3 (May 1, 1968): 392–406.

Dana, Samuel L. *A Muck Manual for Farmers*. Lowell, MA: Bixby and Whiting, 1843.

Davis, Keith F. "'A Terrible Distinctness': Photography of the Civil War Era." In *Photography in Nineteenth-Century America*, edited by Martha A. Sandweiss, 130–79. Fort Worth: Amon Carter Museum, 1991.

"Devil's Dust." *The Spectator*, March 12, 1842.

Dewsbury: The Official Handbook. London: Ed. J. Barrow and Co. Ltd., 1951.

Dickens, Charles. *Sketches by Boz*. London: John Macrone, 1836.

Diderot, Denis, and Jean Le Rond d'Alembert. *Encyclopédie, ou Dictionnaire Raisonné*. Vol. 5. Paris: Briasson David, 1767.

Didi-Huberman, Georges. *Images in Spite of All: Four Photographs from Auschwitz*. Translated by Shane B Lillis. Chicago: University of Chicago Press, 2012.

Didi-Huberman, Georges. "The Index of the Absent Wound (Monograph on a Stain)." Translated by Thomas Repensek. *October* 29 (1984): 63–81.

"The Disinfection of Rags." *New York Times*, December 18, 1884.

Disraeli, Benjamin. *Sybil, or The Two Nations*. London: Oxford University Press, 1975.

"Dissenters, Religious Duty and Religious Practices." *Common Sense, or Every-Body's Magazine* 2, no. 5 (May 1843): 110–14.

"The Dream of the Army Contractor." *Vanity Fair*, August 17, 1861, 77.

Druitt, Robert. *The Principles and Practice of Modern Surgery*. Philadelphia: Lea and Blanchard, 1847.

Edgerton, David. "From Innovation to Use: Ten Eclectic Theses on the Historiography of Technology." *History and Technology* 16, no. 2 (January 1, 1999): 111–36.

Edgerton, David. *The Shock of the Old: Technology and Global History since 1900*. Oxford: Oxford University Press, 2006.

Engels, Friedrich. *The Condition of the Working-Class in England in 1844*. With a Preface Written in 1892. Translated by Florence Kelley Wischnewetzky. London: Allen and Unwin, 1892.

Farrant, Laura, Stig Irving Olsen, and Arne Wangel. "Environmental Benefits from Reusing Clothes." *International Journal of Life Cycle Assessment* 15, no. 7 (2010): 726–36.

Fenn, Elizabeth. *Pox Americana: The Great Smallpox Epidemic of 1775–82*. New York: Hill & Wang, 2001.

"Filthy Imported Rags." *New York Times*, August 2, 1884.

Fletcher, Kate. *Sustainable Fashion and Textiles: Design Journeys*. London: Routledge, 2013.

Fletcher, Kate, and Lynda Grose. *Fashion and Sustainability: Design for Change*. London: Lawrence King, 2012.

Fontaine, Laurence, ed. *Alternative Exchanges: Second-Hand Circulations from the Sixteenth Century to the Present*. New York: Berghahn Books, 2008.

Foster, John Bellamy. *Marx's Ecology: Materialism and Nature*. New York: Monthly Review Press, 2000.

Frankfurter, Felix. *Mr. Justice Holmes and the Constitution: A Review of His Twenty-Five Years on the Supreme Court*. Cambridge, MA: Dunster House Bookshop, 1927.

Frassanito, William A. *Early Photography at Gettysburg*. Gettysburg, PA: Thomas Publications, 1995.

Frassanito, William A. *Gettysburg: A Journey in Time*. New York: Scribner, 1975

Frederickson, George. *The Inner Civil War: Northern Intellectuals and the Crisis of the Union*. New York: Harper & Row, 1965.

Gabrys, Jennifer. "Shipping and Receiving: The Social Death of Electronics." In *Aesthetic Fatigue: Modernity and the Language of Waste*, edited by John Scanlan and John Clark, 274–97. Newcastle upon Tyne: Cambridge Scholars Publishing, 2013.

Garrison, Felding H. "Fracastorius, Athanasius Kircher, and the Germ Theory of Disease." *Science* 31, no. 796 (April 1, 2010): 500–502.

Geller, Jay. *The Other Jewish Question: Identifying the Jew and Making Sense of Modernity*. New York: Fordham University Press, 2011.

Gernsheim, Helmut. *A Concise History of Photography*. New York: Dover, 1986.

Ginsburg, Madeleine. "Rags to Riches: The Second-Hand Clothes Trade, 1700–1978." *Costume: The Journal of the Costume Society* 14 (1980): 121–35.

Glanz, Rudolf. *The Jew in the Old American Folklore*. New York: Waldon, 1961.

Goddard, Nicholas. "'A Mine of Wealth'? The Victorians and the Agricultural Value of Sewage." *Journal of Historical Geography* 22, no. 3 (July 1996): 274–90.

Grant, Susan-Mary. *Oliver Wendell Holmes Jr.: Civil War Soldier, Supreme Court Justice*. New York: Routledge, 2015.

"A Great Social Problem." *Continental Monthly* 6, no. 5 (October 1864): 441–44.

Gregson, Nicky, and Louise Crewe. *Second-hand Cultures*. London: Berg, 2003.

Green, Jonathon. *Cassell's Dictionary of Slang*. London: Wiedenfeld & Nicolson, 2005.

Griffin, Emma. *A Short History of the British Industrial Revolution*. London: Palgrave Macmillan, 2010.

Haggblade, Steven. "The Flip Side of Fashion: Used Clothing Exports to the Third World." *Journal of Development Studies* 26, no. 3 (1990): 505–21.

Hague, D. C. "The Man-Made Fibers Industry." In *The Structure of British Industry*, edited by Duncan Lyall Burn, 2:259–90. Cambridge: Cambridge University Press, 1958.

Haigh, Malcolm H. *The History of Batley, 1800–1974*. Batley, UK: M. H. Haigh, 1978.

Handlin, Oscar, and Mary F. Handlin. "A Century of Jewish Immigration to the United States." Edited by Harry Schneiderman and Morris Fine. *American Jewish Yearbook* 50 (1948–1949): 1–84.

Hansen, Karen Tranberg. "Helping or Hindering? Controversies around the International Second-Hand Clothing Trade." *Anthropology Today* 20, no. 4 (2004): 3–9.

Harley, C. Knick. "Trade: Discovery, Mercantilism and Technology." In *The Cambridge Economic History of Modern Britain*, edited by Roderick Floud and Paul Johnson, 175–203. Cambridge: Cambridge University Press, 2004.

Harris, John M. "'Truthful as the Record of Heaven': The Battle of Antietam and the Birth of Photojournalism." *Southern Cultures* 19, no. 3 (2013): 79–94.

Hayes, John L. *The Fleece and the Loom: Address before the National Association of Wool Manufacturers at the First Annual Meeting in Philadelphia, Sept. 6, 1865*. Boston: Press of John Wilson and Sons, 1865.

Haythornthwaite, Philip J. *Uniforms of the American Civil War, 1861–65*. London: Blandford Press, 1975.

Hayward, Ida M., and George Claridge Druce. *The Adventive Flora of Tweedside*. Arbroath, UK: T. Buncle & Co, 1919.

Head, George. *A Home Tour through the Manufacturing Districts of England, in the Summer of 1835*. J. Murray, 1836.

Hearn, Chester G. *When the Devil Came Down to Dixie: Ben Butler in New Orleans*. Baton Rouge: Louisiana State University Press, 1997.

Heaton, Herbert. *The Yorkshire Woollen and Worsted Industries, from the Earliest Times up to the Industrial Revolution*. Oxford: Clarendon Press, 1966.

Heaton, Herbert, and the Thoresby Society. *Yorkshire Cloth-Traders in the Unit-*

ed States, 1770–1840. Leeds: Thoresby Society, 1944.

Henderson, James. "Industrial Legislation." In *Great Industries of Great Britain*, 2:273–76. London: Cassell, Petter, Galpin, and Co., 1884.

Herbert, Christopher. *Culture and Anomie: Ethnographic Imagination in the Nineteenth Century*. Chicago: University of Chicago Press, 1991.

Hingham, John. *Strangers in the Land: Patterns of American Nativism, 1860–1925*. New Brunswick, NJ: Rutgers University Press, 1955.

Hirst, Francis Wrigley, ed. *Free Trade and Other Fundamental Doctrines of the Manchester School*. London: Harper, 1903.

Hodges, Nathaniel. *Loimologia: Or, an Historical Account of the Plague in London in 1665*. Translated by John Quincy. London: Printed for E. Bell and J. Osborn, 1721.

Hollander, Anne. *Seeing through Clothes*. New York: Viking Press, 1978.

Holmes, Oliver Wendell, Sr. "The Age of Photography." *Atlantic Monthly*, June 1859.

Holmes, Oliver Wendell, Sr. "The Contagiousness of Puerperal Fever," *New England Quarterly Journal of Medicine and Surgery* 1 (1843): 503–30.

Holmes, Oliver Wendell, Sr. "Doings of the Sunbeam." *Atlantic Monthly* 12, no. 69 (July 1863): 1–15.

Holmes, Oliver Wendell, Sr. "The Stereoscope and the Stereograph." In *Soundings from the Atlantic*, 124–65. Boston: Ticknor and Fields, 1864.

Holmes, Oliver Wendell, Jr. *The Dissenting Opinions of Mr. Justice Holmes*. Edited by Alfred Lief. Getzville, NY: Wm. S. Hein Publishing, 1929.

Hotten, John C. *The Slang Dictionary*. London: John Camden Hotten, 1869.

Hudson, Pat. *The Genesis of Industrial Capital: A Study of West Riding Wool Textile Industry, c. 1750–1850*. Cambridge: Cambridge University Press, 1986.

Innis, Thomas. *The Skin, in Health and Disease: A Concise Manual*. London: Whittaker, 1849.

Jenkins, D. T. "The Western Wool Textile Industry in the Nineteenth Century." In *The Cambridge History of Western Textiles*, edited by D. T. Jenkins. Cambridge: Cambridge University Press, 2003.

Jenkins, D. T., and Kenneth G. Ponting. *The British Wool Textile Industry, 1770–1914*. London: Ashgate, 1982.

Johnson, Steven. *The Ghost Map: The Story of London's Most Terrifying Epidemic—and How It Changed Science, Cities, and the Modern World*. New York: Riverhead Books, 2006.

Jones, Ann Rosalind, and Peter Stallybrass. *Renaissance Clothing and the Materials of Memory*. Cambridge: Cambridge University Press, 2001.

Jørgensen, Finn Arne. *Making a Green Machine: The Infrastructure of Beverage Container Recycling*. New Brunswick, NJ: Rutgers University Press, 2011.

Jubb, Samuel. *The History of the Shoddy-Trade: Its Rise, Progress, and Present Position*. London: Houlston and Wright, 1860.

Kaplan, Fred. *Thomas Carlyle: A Biography*. Ithaca, NY: Cornell University Press, 1983.

"Killing Disease Germs: Another Forward Step in Rag Disinfection." *New York Times*, October 17, 1885.

Kittredge, Henry G. "Shoddy: Or the History of a Woolen Rag." *Technology Quarterly* 19, no. 2 (1906): 65–82.

Krauss, Rosalind E. "Notes on the Index: Part 1." In *The Originality of the Avant-Garde and Other Modernist Myths*. Cambridge, MA: MIT Press, 1991.

Kraut, Alan M. *Silent Travelers: Germs, Genes, and the "Immigrant Menace."* New York: Basic Books, 1994.

Lambert, Miles. "Cast-off Wearing Apparell." *Textile History* 35, no. 1 (2004): 1–26.

Landes, David S. *The Unbound Prometheus*. Cambridge: Cambridge University Press, 1969.

Law, Edwin. "The Discovery and Early History of the Shoddy and Mungo Trades." *Batley Reporter*, November 13, 1880.

Lee, Anthony W., and Elizabeth Young. *On Alexander Gardner's Photographic Sketch Book of the Civil War*. Berkeley: University of California Press, 2007.

Leja, Michael. *Looking Askance: Skepticism and American Art from Eakins to Duchamp*. Berkeley: University of California Press, 2004.

Lemire, Beverly. "Consumerism in Preindustrial and Early Industrial England: The Trade in Secondhand Clothes." *Journal of British Studies* 27, no. 1 (1988): 1–24.

Lemire, Beverly. "Peddling Fashion: Salesmen, Pawnbrokers, Taylors, Thieves and the Second-Hand Clothes Trade in England, c. 1700–1800." *Textile History* 22, no. 1 (1991): 67–82.

Lemire, Beverly. "Shifting Currency: The Culture and Economy of the Second Hand Trade in England, c. 1600–1850." In *Old Clothes, New Looks: Second Hand Fashion*, edited by Alexandra Palmer and Hazel Clark, 29–47. Oxford: Berg Publishers, 2005.

"Letter from October 31, 1861 from Davies to Brig.-Gen. L. Thomas." *New York Times*, November 2, 1861.

"Letter from the Rev. R. S. Bailey to Lord Brougham." *Common Sense, or Everybody's Magazine*, January 1843.

Lévi-Strauss, Claude. *La pensée sauvage*. Paris: Plon, 1962.

Linebaugh, Peter. *The London Hanged: Crime and Civil Society in the Eighteenth Century*. London: Verso, 2006.

Lister, Joseph. "On the Effects of the Antiseptic System of Treatment upon the Salubrity of a Surgical Hospital, Part 1." *Lancet* 95, no. 2419 (1870): 40–42.

Livermore, Mary A. *My Story of the War: A Woman's Narrative*. Hartford, CT: A. D. Worthington and Co., 1896.

Lock, Charles G. Warnford, ed. *Spons' Encyclopaedia of the Industrial Arts, Manufactures, and Raw Commercial Products*. London: E. and F. N. Spon, 1882.

Macpherson, John. *Epidemic Cholera: Its Mission and Mystery, Haunts and Havocs, Pathology and Treatment: With Remarks on the Question of Contagion, the Influence of Fear, and Hurried and Delayed Interments*. New York: Carleton, 1866.

Magee, Gary Bryan. *Productivity and Performance in the Paper Industry: Labour, Capital, and Technology in Britain and America, 1860–1914*. Cambridge: Cambridge University Press, 1997.

Männistö-Funk, Tiina. "The Crossroads of Technology and Tradition: Vernacular Bicycles in Rural Finland, 1880–1910." *Technology and Culture* 52, no. 4 (2011): 733–56.

Markel, Howard. *Quarantine!: Eastern European Jewish Immigrants and the New York City Epidemic of 1892*. Baltimore: Johns Hopkins University Press, 1997.

Marx, Karl. *Capital*. Vol. 1. Trans. Ben Fowkes. London: Penguin Classics, 1990.

Marx, Karl. *The Eighteenth Brumaire of Louis Bonaparte*. Trans. Daniel De Leon. New York: International Publishers, 1898.

Marx, Karl, and Friedrich Engels. *Collected Works of Marx and Engels*. Vol. 14, *1855–1856*. New York: International Publishers, 1980.

Marx, Karl, and Friedrich Engels. *The German Ideology*. London: Lawrence & Wishart, 1965.

Marx, Leo. "The Idea of 'Technology' and Postmodern Pessimism." In *Does Tech-*

nology Drive History? The Dilemma of Technological Determinism, edited by Leo Marx and Merritt Roe Smith, 237–58. Cambridge, MA: MIT Press, 1994.

Marx, Leo. *The Machine in the Garden: Technology and the Pastoral Ideal in America*. New York: Oxford University Press, 1964.

Marx, Leo. "Technology: The Emergence of a Hazardous Concept." *Social Research* 64, no. 3 (Fall 1997): 965–88.

Mauersberger, Herbert R. *Matthews' Textile Fibers: Their Physical, Microscopic, and Chemical Properties*. London: Chapman and Hall, 1954.

Mayhew, Henry. *London Labour and the London Poor*. London: G. Woodfall, 1851.

McCulloch, J. R. *A Dictionary, Practical, Theoretical, and Historical, of Commerce and Commercial Navigation*. London: Longman, Brown, Green, and Longmans, 1850.

McGaw, Judith A. *Most Wonderful Machine: Mechanization and Social Change in Berkshire Paper Making, 1801–1855*. Princeton, NJ: Princeton University Press, 1987.

McNeill, William H. *Plagues and Peoples*. New York: Doubleday, 1977.

Megraw, Robert H. *Textiles and the Origin of Their Names*. New York: n.p., 1906.

Meltzer, Milton. *Walt Whitman: A Biography*. Brookfield, CT: Twenty-First Century Books, 2002.

Mendelsohn, Adam. *The Rag Race: How Jews Sewed Their Way to Success in America and the British Empire*. New York: NYU Press, 2016.

Mnookin, Jennifer. "The Image of Truth: Photographic Evidence and the Power of Analogy." *Yale Journal of Law & the Humanities* 10, no. 1 (May 8, 2013): 1–74.

Moor, Nigel. "Batley at Work: The Rise and Fall of a Textile Town," Employment & Industry Programme Discussion Paper No. 1 (Nigel Moor and Associations for the Batley Community Development Project, December 1974).

Morford, Henry. *The Days of Shoddy: A Novel of the Great Rebellion in 1861*. Philadelphia: T. B. Peterson & Brothers, 1863.

Morley, John. *The Life of Richard Cobden*. Vol. 1. London: Fisher Unwin, 1896.

Morrell, J. B. "Wissenschaft in Worstedopolis: Public Science in Bradford, 1800–1850." *British Journal for the History of Science* 18, no. 1 (March 1985): 1–23.

"Mr. Shoddy Having Made Much Money." *Dollar Monthly Magazine*, March 19, 1864.

Mudge, Enoch Redington, and John Lord Hayes. *Report upon Wool and Manufactures of Wool*. Washington, DC: US Government Printing Office, 1868.

National Association of Wool Manufacturers. "Statistics of the British Wool Manufacture." *Bulletin of the National Association of Wool Manufacturers* 1, no. 1 (January 1869): 43–58.

Nevins, Allan. *The War for the Union: The Improvised War, 1861-1862*. Vol. 1. New York: Charles Scribner's, 1959.

Nevins, Allan. *The War for the Union: The Organized War, 1863–1864*. Vol. 3. New York: Charles Scribner's, 1971.

Nohl, Johannes. *The Black Death*. London: Allen & Unwin, 1926.

Norris, Lucy. "Economies of Moral Fibre? Materializing the Ambiguities of Recycling Charity Clothing into Emergency Aid Blankets." *Journal of Material Culture* 17, no. 4 (December 2012): 389–404.

"Notes of the Rebellion: Shabby Uniforms." *New York Times*, July 20, 1861.

"The Odyssey of Some Australian Weeds Found in an English Field." *Illustrated London News*, November 3, 1951.

Olcott, Henry S. "Song of the Shoddy." *Vanity Fair*, September 21, 1861, 142.

Oliver, Thomas, ed. *Dangerous Trades: The Historical, Social, and Legal Aspects of Industrial Occupations as Affecting Health*. London: John Murray, 1902.

"Only Americans Wear Virgin Wool: Order Gives Our Soldiers Pure Worsted Uniforms, Though 'Shoddy' Satisfies Fighters of Other Nations." *New York Times*, February 10, 1918.

Orvell, Miles. *The Real Thing: Imitation and Authenticity in American Culture, 1880–1940*. Chapel Hill: University of North Carolina Press, 1989.

Parker, Edwin Brewington. *Final Report of United States Liquidation Commission, War Department*. Washington, DC: US Government Printing Office, 1920.

Parliamentary Debates, House of Commons Debates, 24 February 1842. Vol. 60, cols. 1018–82.

"Patent Rag Shredding Machines." *Textile Manufacturer* 44 (August 15, 1918): 225–26.

Peirce, Charles S. *Philosophical Writings of Peirce*. New York: Dover, 1961.

Pickering, Andrew. *The Mangle of Practice: Time, Agency, and Science*. Chicago: University of Chicago Press, 1995.

Polanyi, Michael. *The Tacit Dimension*. Chicago: University of Chicago Press, 1966.

"Preparing Old Woolen Rags for Shoddy Clothing." *Scientific American* 75, no. 3 (July 18, 1896): 37.

Price, Laura. "Immigrants and Apprentices: Solutions to the Post-War Labour Shortage in the West Yorkshire Wool Textile Industry, 1945–1980." *Textile History* 45, no. 1 (May 2014): 32–48.

Quain, Richard. *A Dictionary of Medicine: Including General Pathology, General Therapeutics, Hygiene, and the Diseases of Women and Children.* London: Longmans, Green and Company, 1882.

Radcliffe, J. W. *Woollen and Worsted Yarn Manufacture.* Manchester: Emmott & Co., 1953.

"The Rag Trade." *New York Times,* July 30, 1871.

Rainnie, G. F. "The Woollen and Worsted Industry." In *The Structure of British Industry,* edited by Duncan Lyall Burn, 222–58. Cambridge: Cambridge University Press, 1958.

Reach, Angus B. *Fabrics, Filth and Fairy Tents: The Yorkshire Textile Districts in 1849.* Edited by C. Aspin. Hebden Bridge, UK: Royd Press, 2007.

Reed, Isaac George. *The Russian Ball, or The Adventures of Miss Clementina Shoddy, a Humorous Description in Verse.* New York: Carleton, 1863.

Reynard, Pierre-Claude. "Unreliable Mills: Maintenance Practices in Early Modern Papermaking." *Technology and Culture* 40, no. 2 (April 1999): 237–62.

Risch, Erna. *Quartermaster Support of the Army: A History of the Corps, 1775–1939.* Washington, DC: US Army, Center of Military History, 1989.

Rosenberg, Charles E. *The Cholera Years: The United States in 1832, 1849, and 1866.* Chicago: University of Chicago Press, 1962.

Sandweiss, Martha A., ed. *Photography in Nineteenth-Century America.* New York: Abrams, 1991.

Scanlan, J. "In Deadly Time: The Lasting on of Waste in Mayhew's London." *Time and Society* 16, no. 2–3 (2007): 189–206.

Scanlan, John. *On Garbage.* London: Reaktion, 2005.

Secondhand (Pepe). Directed by Hanna Rose Shell and Vanessa Bertozzi. Boston: Fabrik Films, distributed by Third World Newsreel, 2007.

Shapter, Thomas. *The History of the Cholera in Exeter in 1832.* London: J. Churchill, 1849.

Shaw, Madelyn. *Homefront and Battlefield: Quilts and Context in the Civil War.* Lowell, MA: American Textile History Museum, 2012.

Shaw, Madelyn. "Slave Cloth and Clothing Slaves: Craftsmanship, Commerce, and Industry." *Journal of Early Southern Decorative Arts,* 2012.

Shell, Hanna Rose. "A Global History of Secondhand Clothing." *Spreadable Me-*

dia. 2013. http://spreadablemedia.org/essays/shell/#.XQu1dehKiUl.

Shell, Hanna Rose. "Shoddy Heap: A Material History between Waste and Manufacture." *History and Technology* 30, no. 4 (October 2, 2014): 374–94.

Shell, Hanna Rose. "Textile Skin." *Transition* 13, no. 96 (2006): 152–63.

"Shoddy." *Mechanics Magazine: American Artisans and Patent Record*, November 27, 1867.

"Shoddy: Its Material and Manufacture." *Scientific American* 15, no. 13 (September 22, 1866).

"Shoddy: The Way It Is Made." *Scientific American* 5, no. 15 (October 12, 1861).

"Shoddy Bedding." *Annals of Hygiene* 11, no. 2 (1896): 125–26.

The Shoddy Industry vs. the Virgin Wool Industry. Chicago: National Sheep and Wool Bureau, 1919.

"*Shoddy vs. Pure Wool*: Hearings before the Committee on Ways and Means, House of Representatives, Fifty-Seventh Congress, First Session," Pub. L. No. 413, SD (Washington, DC: Government Printing Office, 1902).

Simmonds, Peter Lund. *Waste Products and Undeveloped Substances; or, Hints for Enterprise in Neglected Fields*. London: R. Hardwicke, 1862.

Simmonds, Peter Lund. *Waste Products and Undeveloped Substances: A Synopsis of Progress Made in Their Economic Utilisation during the Last Quarter of a Century at Home and Abroad*. London: R. Hardwicke, 1873.

Simon, Sir John, and the Great Britain Privy Council. *Public Health: Eighth Report of the Medical Officer of the Privy Council, with Appendix, 1865*. London: George E. Eyre and William Spottiswoode, 1866.

Smith, William M. "Contagious Diseases Propagated by Rags, and the Necessity of Disinfection." *The Sanitarian* 15, no. 193 (December 1, 1885): 481–524.

Snow, John. "Cholera and the Water Supply in the South Districts of London in 1854." *The Times*, June 26, 1856.

Snow, John. *On the Mode of Communication of Cholera*. London: John Churchill, 1855.

Snowden, Frank. *Naples in the Time of Cholera, 1884–1911*. Cambridge: Cambridge University Press, 1995.

Sollors, Werner. *Beyond Ethnicity*. Oxford: Oxford University Press, 1986.

Sontag, Susan. *On Photography*. New York: Anchor Books, 1990.

"Specifications on Patents Recently Filed." *Mechanics Magazine* 66 (January 3, 1857): 16.

Stallybrass, Peter. "Marx's Coat." In *Border Fetishisms: Material Objects in*

Unstable Spaces, edited by Patricia Spyer, 182–207. New York: Routledge, 1998.

"Statute Prohibiting Use of Shoddy in Manufacture of Comfortables Unconstitutional." *Business Law Journal* 8 (1926): 128–29.

Sternberg, George Miller. *Disinfection and Individual Prophylaxis against Infectious Diseases*. Concord, NH: Republican Press Association, 1886.

Sternberg, George M., et al. "Report of the Committee on Disinfection of Rags." In *Public Health Papers and Reports*, 12:170–97. Concord, NH: Republican Press Association, 1887.

Stobart, Jon, and Ilja Van Damme, eds. *Modernity and the Second-Hand Trade: European Consumption Cultures and Practices, 1700–1900*. New York: Palgrave Macmillan, 2011.

Stowell, Sheila. "Teufelsdröckh as Devil's Dust." *Carlyle Newsletter*, no. 9 (Spring 1988): 31–33.

Strasser, Susan. *Waste and Want: A Social History of Trash*. New York: Henry Holt, 1999.

Stuart, John A. E. *The Literary Shrines of Yorkshire*. London: Longmans, Green & Co., 1892.

Stuart, John A. E. "Rags and Their Products in Relation to Health." In *Dangerous Trades: The Historical, Social, and Legal Aspects of Industrial Occupations as Affecting Health*, edited by Thomas Oliver, 644–47. London: John Murray, 1902.

Supervising Surgeon General. *Operations of the United States Marine-Hospital Service in 1895*. Washington, DC: US Government Printing Office, 1896.

Sutter, Paul. "Tropical Conquest and the Rise of the Environmental Management State: The Case of U.S. Sanitary Efforts in Panama." In *Colonial Crucible: Empire in the Making of the Modern American State*, edited by Alfred W. McCoy and Francisco A. Scarano, 317–26, 605–6. Madison: University of Wisconsin Press, 2009.

Sykes, J. F. J. *Public Health Problems*. London: Charles Scribner's Sons, 1892.

Taussig, F. W. *The Tariff History of the United States*. New York: G. P. Putnam's Sons, 1910.

Tennyson, G. B. *Sartor Called Resartus: The Genesis, Structure, and Style of Thomas Carlyle's First Major Work*. Princeton, NJ: Princeton University Press, 1966.

Thompson, E. P. *The Making of the English Working Class*. New York: Random House, 1963.

Tomes, Robert. "The Fortunes of War: How They Are Made and Spent." *Harper's Monthly*, July 1864.

Trachtenberg, Alan. "Albums of War: On Reading Civil War Photographs." *Representations* 9 (Winter 1985): 1–32.

Trachtenberg, Alan. *Reading American Photographs: Images as History, Mathew Brady to Walker Evans*. New York: Noonday Press, 1990.

The Truth in Reworked Wool. New York: Atlas Publishing, 1920.

Tucker, Jennifer. *Nature Exposed: Photography as Eyewitness in Victorian Science*. Baltimore: Johns Hopkins University Press, 2014.

Twain, Mark, and Charles Dudley Warner. *The Gilded Age: A Tale of Today*. Hartford, CT: American Publishing Company, 1873.

"Use of Shoddy Is Greatest in America: Workingmen Here Literally Wearing the World's Old Clothes." *New York Times*, July 10, 1904, financial supplement, 4.

Vinten-Johansen, Peter, et al. *Cholera, Chloroform, and the Science of Medicine: A Life of John Snow*. Oxford: Oxford University Press, 2003.

Von Bergen, Werner, and Walter Krauss. *Textile Fiber Atlas: A Collection of Photomicrographs of Old and New Textile Fibers*. New York: Textile Book Publishers, 1949.

Weigley, Russell Frank. *Quartermaster General of the Union Army: A Biography of M. C. Meigs*. New York: Columbia University Press, 1959.

"What Next?" *New York Times*, July 26, 1861.

Whipple, E. P. "Shoddy," *Atlantic* 27, no. 161 (March 1871): 338–48.

Whipple, E. P. *Success and Its Conditions*. Boston: James R. Osgood, 1871.

White, G. Edward. *Justice Oliver Wendell Holmes: Law and the Inner Self*. Oxford: Oxford University Press, 1995.

Wild, J. P. "The Romans in the West: 600 BC–AD 400." In *The Cambridge History of Western Textiles*, edited by D. T. Jenkins. Cambridge: Cambridge University Press, 2003.

Wilson, Christopher Kent. "Marks of Honor and Death: *Sharpshooter* and the Peninsular Campaign of 1862." In *Winslow Homer: Paintings of the Civil War*, edited by Marc Simpson, 25–42. San Francisco: Fine Arts Museums of San Francisco/Bedford Arts, 1988.

Wilson, Elizabeth. *Adorned in Dreams: Fashion and Modernity*. New Brunswick, NJ: Rutgers University Press, 2003.

Wilson, Mark R. *The Business of Civil War: Military Mobilization and the State, 1861–1865*. Baltimore: Johns Hopkins University Press, 2006.

Winter, James. *Secure from Rash Assault: Sustaining the Victorian Environment*. Berkeley: University of California Press, 1999.

Withington, Charles Francis. "Transmission of Infectious Diseases through the Medium of Rags." In *Eighteenth Annual Report of the Massachusetts State Board of Health*, 3–69, 1886.

"The Woollen and Worsted Trade of Great Britain," *Bulletin of the National Association of Wool Manufacturers* 1, no. 1 (1869): 46–59.

"Yorkshire." *Westminster Review* 71, no. 140 (April 1859): 179–95.

Zakim, Michael. *Ready-Made Democracy: A History of Men's Dress in the American Republic, 1760–1860*. Chicago: University of Chicago Press, 2003.

Zimring, Carl. *Aluminum Upcycled: Sustainable Design in Historical Perspective*. Baltimore: Johns Hopkins University Press, 2017.

Zimring, Carl. *Cash for Your Trash: Scrap Recycling in America*. New Brunswick, NJ: Rutgers University Press, 2009.

Zimring, Carl A., and William L. Rathje, eds. *Encyclopedia of Consumption and Waste: The Social Science of Garbage*. London: SAGE Publications, 2012.

Index

Pages in italics refer to illustrations.